The Foundations of Causal Decision Theory

This book defends the view that any adequate account of rational decision making must take a decision maker's beliefs about causal relations into account.

The early chapters of the book introduce the nonspecialist to the rudiments of expected utility theory. The major technical advance offered by the book is a "representation theorem" that shows that both causal decision theory and its main rival, Richard Jeffrey's logic of decision, are instances of a more general conditional decision theory. The book solves a long-standing problem for Jeffrey's theory by showing for the first time how to obtain a unique utility and probability representation for preferences and judgments of comparative likelihood. The book also contains a major new discussion of what it means to suppose that some event occurs or that some proposition is true.

In providing the most complete and robust defense of causal decision theory yet presented, the book will be of interest to a broad range of readers in philosophy, economics, psychology, mathematics, and artificial intelligence.

James M. Joyce is Associate Professor of Philosophy at the University of Michigan.

T0275707

**Cambridge Studies in Probability,
Induction, and Decision Theory**

The Foundations of Causal Decision Theory

James M. Joyce

CAMBRIDGE
UNIVERSITY PRESS

PUBLISHED BY THE PRESS SYNDICATE OF THE UNIVERSITY OF CAMBRIDGE
The Pitt Building, Trumpington Street, Cambridge, United Kingdom

CAMBRIDGE UNIVERSITY PRESS
The Edinburgh Building, Cambridge CB2 2RU, UK   http: //www.cup.cam.ac.uk
40 West 20th Street, New York, NY 10011-4211, USA   http://www.cup.org
10 Stamford Road, Oakleigh, Melbourne 3166, Australia

First published 1999

Typeface Times 10/12 pt.   *System* PageMaker [BTS]

A catalog record for this book is available from the British Library.

Library of Congress Cataloging-in-Publication Data
Joyce, James M., 1958–
The foundations of causal decision theory / James M. Joyce.
        p.   cm.
Includes bibliographical references and index.
ISBN 0-521-64164-0 (hbk)
1. Decision making.   I. Title.
BD184.J68   1999
128'.3 – dc21
                                                              98-36028
                                                              CIP

ISBN 0-521-64164-0 hardback

Transferred to digital printing 2004

For my wife, Emily; my son, Jacob;
and my parents, John and Margaret.

Contents

Preface

This book is about how choices should be made. It espouses a version of the *expected utility hypothesis*. This doctrine, which is widely endorsed by economists, philosophers, and others interested in normative accounts of choice, enshrines subjective expected utility maximization as the central principle of rational decision making. The particular brand of the expected utility hypothesis to be defended here is called *causal decision theory*. Causal decision theorists maintain that it is impossible to make sense of rational agency without giving due consideration to the agent's beliefs about what his or her actions are likely to causally promote, and moreover that it is impossible to understand these "causal" beliefs as ordinary beliefs about noncausal propositions. While I will be defending the causal theory in this work, my main objective is to develop its formalism more completely than it has heretofore been developed, and to use the results to illuminate the nature of causal beliefs and their role in rational choice. One main lesson of these investigations is that there is less difference than is usually thought between causal decision theory and its main competitor, "evidential" decision theory, at least as far as the deep foundational issues are concerned. A second lesson is that causal beliefs are somewhat more tractable to formal analysis than is usually alleged.

There are a great number of people who have helped me with the work presented here. Allan Gibbard and Larry Sklar were in on the project from the start, and both have given me valuable advice and support all along the way. I have also received useful input from Elizabeth Anderson, Brad Armendt, Bob Batterman, David Christensen, Justin D'Arms, John Devlin, Alan Hajek, Dan Farrell, Mark Giuffrida, Bill Harper, Sally Haslanger, Chris Hill, Chris Hitchcock, Dick Jeffrey, Mark Kaplan, Paul Killey, Robert Mabrito, John Muckler, Gerhard Nuffer, Peter Railton, Gideon Rosen, Brian Skyrms, Bas van Fraassen, Peter Vallentyne, David Velleman, Peter Vranas, Ken Walton, Scott Wilderman, Mark Wilson, Steve Yablo, Ken Zasadny, and Lyle Zynda. Special thanks also go to Terry Smith, who helped me through some of the early stages of this project, and to Louis Loeb, whose wise advice toward the end (sometimes not sufficiently heeded) helped me see the forest for the trees. Thanks also to Terry Moore, Gwen Seznec, and Susan Thornton of Cambridge University Press for making the publication process so easy.

It will be obvious to anyone who reads this work that my two greatest intellectual debts are to Richard Jeffrey and Allan Gibbard. Jeffrey's work, particularly his magnificent *Logic of Decision*, opened my eyes to expected utility theory and caused me to see that his version of it must contain an important kernel of the truth. Gibbard's uncommonly clear-headed insights about rational choice and causation helped me see what was really going on in the dispute between causal and evidential decision theorists, and to recognize that the causal view must also contain an important kernel of the truth. Brad Armendt and J. Howard Sobel also deserve thanks. Armendt's idea to formulate causal decision theory in terms of conditional preferences was the catalyst for the representation theorem in Chapter 7. Sobel's papers on "partition problems" in causal decision theory led me to appreciate the importance of the property of partition invariance in expected utility theory. This turned out to be a significant turning point in my thinking since it led me to appreciate the crucial point that Jeffrey's theory could be right as a theory of value even if it was not, ultimately, a "logic of decision."

My greatest personal debts are to my wife, Emily Santer, without whose support I would never have been able to develop any of these ideas, and to the members of my Department at the University of Michigan, whose unflagging support gave me the confidence to produce this work.

Introduction
A Chance to Reconsider

So, you have decided to read this book. Are you sure that's wise? Reading a tome entitled *The Foundations of Causal Decision Theory* is a sure sign of interest in the study of rational choice, which is a very unfortunate interest to have. You probably have heard all the jokes about decision theorists who cannot make wise choices. (How many does it take to screw in a light bulb? Who knows; you can't draw up a decision matrix in the dark. What do you call a decision theorist with a bank balance? A meal ticket.) Unfortunately, there is more than a grain of truth in these jests. It is widely known, at least among people who have no interest in rational choice theory, that about half of those who study decision making are patsies who can be easily exploited. Here are some facts of which you may be unaware: Most business schools keep a few decision theorists on staff (and overpay them grossly) to make it easier to win money at the weekly poker game. Ninety percent of all swampland is owned by people who have read Leonard Savage's *The Foundations of Statistics*. In most banks one can get a loan without putting up collateral merely by showing that one has a product that appeals to *Econometrica* subscribers. No one has ever paid the full sticker price for a car who did not know that Daniel Bernoulli invented the concept of utility. What P. T. Barnum really said was that there is an *expected utility* theorist born every minute.

You've never heard of any of this? I'm not a bit surprised. It has been common practice in all cultures at all times to identify people with an interest in decision theory at an early age, and to keep them in the dark about their plight so as not to spoil the chances of easy exploitation. I am only being permitted to reveal this information now because it has recently been discovered that it does not matter whether or not a person realizes that he is a sucker. The tendency to make foolish choices has long been known to run in families. Until recently this was thought to be a matter of "nurture" rather than "nature." The teaching of sound decision-theoretic principles was believed to engender unsound decision making practices in one out of every two cases. Parents leaving copies of *The Theory of Games and*

1

Economic Behavior open for impressionable eyes to see, cautioning against playing the lottery, delivering stern lectures on the importance of not crying over spilt milk, and the like, were believed somehow to cause their children to buy retail, overestimate their odds in games of chance, and develop a love of decision theory.

Surprisingly, it turns out that being a sucker is congenital. There is a "patsy gene"! It causes its carriers both to have an interest in decision theory and to make lousy choices. If you have this gene, you can resign yourself to a life of misery and exploitation, since that is sure to be your lot. You will never get your money's worth. You will always be an easy mark for unscrupulous operators bent on exploitation. And, if your ship should happen to come in, you can be sure that someone will talk you out of it at the pier. The silver lining in this otherwise dark cloud is that you can enjoy the study of decision theory in peace. It will not make you any better at choosing actions, but it will not make you any worse either.

I am sure you are wondering how to tell whether or not you have the gene. At the moment there is no reliable biological test, so we are going to have to make inferences from your behavior. People with no interest in the theory of rational decision making always lack the gene. You are not one of these, however, since a noncarrier would never even pick up a book entitled *The Foundations of Causal Decision Theory*. Among people like you, who have some interest in rational choice theory, 50 percent have the gene and 50 percent do not. A history of poor decision making is a sure indicator of having the gene. About 49 percent of decision theorists fall into this class. But, this probably is not you. I suspect that you have made fairly wise decisions up to now. So, the good news is that the odds are 50:1 that you are not a carrier. The bad news is that you may be one of the unlucky 1 percent who have the virulent, late-onset form of sucker disease. These are the people who get talked out of their life savings on the telephone, who spend themselves into penury giving money to television preachers, and so on. How can you tell whether this is to be your lot?

As it turns out you are doing the one thing that is known to provide a foolproof test. Research has shown that this very book provides a completely reliable way of determining whether or not a person has the sucker gene. Anyone who chooses to read as far as page 3 has the gene; anyone who stops on page 2 or before lacks it. Research also shows that people who turn to page 3 tend to enjoy reading the book (if only for the fun of detecting all the fallacies), and reading it does not have any deleterious side effects. You are on page 2 now. Think carefully about what you should do next.

You have made a wise choice! Of course, I am sorry for you that you made it because it indicates that your life is going to be miserable. On the bright side, your life would have been miserable whether or not you turned the page, and you will enjoy reading the book (at least a little, for the fallacies). Too bad about the gene, but you cannot change your basic biological makeup and you may as well get as much pleasure as you can out of the bad hand you have been dealt. No use crying over spilt milk, as they say.

You have just faced a *Newcomb problem*. These are choice situations in which one option (e.g., not turning to page 3) reliably indicates the presence of some desirable state of affairs (lacking the patsy gene) *without doing anything to bring that state about*, while another option (turning the page) reliably indicates the presence of some undesirable state of affairs (having the gene) but, again, *without doing anything to bring that state about*. What makes the problem interesting is that this second, less propitious option has benefits not associated with the first (e.g., the pleasure of finding the fallacies). Let us call the first option the *auspicious act* since it serves as a sign or augury of favorable results. The second option will be the *efficacious act* since it plays a causal role in helping to secure the small side benefit. As we shall see later, there are a variety of choice problems with this curious structure.

There are two schools of thought on the issue of how rational agents should behave when faced with Newcomb problems. Proponents of *evidential decision theory* feel that the auspicious option should always be selected. Actions, they believe, ought to be evaluated in terms of the *evidence* they provide for thinking that desirable outcomes will ensue. Defenders of *causal decision theory*, on the other hand, claim that acts are best assessed on the basis of their ability to *causally promote* desirable outcomes. A rational agent, on their view, will always perform an act that is *maximally efficacious* in bringing about desirable consequences. Thus, while an evidential decision theorist would have encouraged you *not* to turn to page 3 of this book (and presumably would not have turned it herself), a causal decision theorist would have advised the opposite.

Both these theories characterize rational desire and choice in terms of *subjective expected utility maximization*. The common ground here is the Bayesian doctrine that the strengths of a rational agent's beliefs can always be measured by a *subjective probability function* P defined over *states of the world*, and the view, which comes down to us from Daniel Bernoulli, David Ricardo, John Stuart Mill, and others, that an agent's desires can be described in terms of a real-valued *utility func-*

tion u defined over *outcomes*. Following Richard Jeffrey, proponents of evidential decision theory maintain that the utility of an act A is best identified with its *news value*, which is given by

Jeffrey's Equation. $V(A) = \sum_S P(S/A)u(O[A, S])$

where S ranges over a set of mutually exclusive and jointly exhaustive "states of the world," where $P(S/A) = P(A \& S)/P(A)$ is the decision maker's *subjective conditional probability* for the state S given A, and where $u(O[A, S])$ is the utility of the outcome that A would produce if it were performed when S obtained. $V(A)$ captures the sense in which A provides the decision maker with evidence for thinking that desirable outcomes will ensue.

Advocates of causal decision theory think that actions should be chosen on the basis of their efficacy value, which is defined using

Stalnaker's Equation. $U(A) = \sum_S P(S \backslash A)u(O[A, S])$

Here $P(S \backslash A)$ is a probability that is supposed to capture the decision maker's beliefs about the extent to which the act A is likely to causally promote the state S. (We will discuss the definition of $P(\bullet \backslash A)$ at length in Chapters 5 and 6.)

In most cases there is no conflict between the evidential and causal approaches to decision making because acts usually indicate good results by *causing them*, which makes U-maximization and V-maximization equivalent. In Newcomb problems, however, indicating and causing come apart, and auspiciousness is no longer a reliable mark of efficacy. There has been a great deal of discussion of the differences between causal and evidential decision theories, and the broad (albeit not universal) consensus is that causal decision theory gets the answers right in situations where the two approaches disagree. It seems clear, for example, that denying oneself the pleasure of finding the fallacies in this book merely to give oneself evidence that one is not a congenital sucker is irrational since one gains nothing at all by doing it.

The difficulty is that from the theoretical point of view causal decision theory is something of a mess; it lacks an appropriate foundation. Evidential decision theory, in contrast, is the model of what a decision theory should be as far as foundational matters are concerned (or so it will be argued). The standard method for justifying any version of expected utility theory involves proving a *representation theorem* that shows that an agent whose beliefs and preferences satisfy certain

4

axiomatically specified constraints will automatically behave as if she is maximizing expected utility as the theory defines it. Such a theorem ensures that the theory's concept of expected utility makes sense and that it can be applied across a broad range of decision situations. It is essential that a decision theory have a representation theorem before it can be taken seriously. As we shall see in Chapter 4, Ethan Bolker has proved a powerful and elegant representation theorem for evidential decision theory that sets the standard by which all other representation theorems should be judged. No similarly compelling result has yet been obtained to serve as a foundation for causal decision theory.

This leaves us in a difficult position. Our best account of rational decision making – the one that seems to give the right answers in Newcomb problems – lacks the minimum theoretical foundation necessary to justify its use, whereas the account that has an adequate theoretical underpinning – the one for which an acceptable representation theorem can be proved – sometimes gives wrong answers. Thus, decision theorists appear to be faced with a choice between an ill-founded theory with true consequences and a well-founded theory with false consequences.

Fortunately, these foundational differences are not as serious as they appear. I will show how to express Jeffrey's Equation and Stalnaker's Equation as instances of a general *conditional expected utility theory* whose defining equation is

$$V(X\|A) = \sum_S \frac{P(S \,\&\, X\|A)}{P(X\|A)} u(O[A, S])$$

where X is any proposition expressible as a disjunction of states, $P(Y\|A)$ is the probability that an agent associates with a proposition Y when she supposes that she will perform A, and $u(O[A, S])$ is her utility for the outcome that would ensue if S were to obtain when A was performed. $V(X\|A)$ is the news value associated with X *on the supposition that A is performed*. I will argue that any tenable theory of rational choice must postulate expected utilities that obey some version of this equation and that it should ask people to maximize, not the unconditional utilities of their acts, but the utilities of their acts *conditional on the supposition that they are performed*. In other words, a rational agent should always choose to perform an act A such that $V(A\|A)$ is greater that $V(B\|B)$ for any alternative B.

To see that this characterization of prudential rationality is broad enough to encompass both the evidential and causal approaches,

5

notice first that by substituting the ordinary conditional probability $P(\bullet/A)$ for $P(\bullet\|A)$ one obtains an "evidentialist" conditional utility theory whose defining equation is

$$V(X/A) = \sum_s \frac{P(S \& X/A)}{P(X/A)} u(O[A, S])$$

Similarly, substituting the causal probability $P(\bullet\backslash A)$ for $P(\bullet\|A)$ yields a notion of conditional expected utility appropriate to causal decision theory:

$$V(X\backslash A) = \sum_s \frac{P(S \& X\backslash A)}{P(X\backslash A)} u(O[A, S])$$

Since these equations reduce to Jeffrey's Equation and Stalnaker's Equation, respectively, when $X = A$, evidential and causal decision theory can each be understood as different versions of a generalized conditional decision theory when the relevant conditional probabilities are properly interpreted.

What we gain by moving to conditional decision theory is a unified framework within which both evidential and causal decision theory can be expressed in formally similar terms. In the penultimate chapter of this book I prove a representation result for conditional decision theory that generalizes Bolker's theorem. Since this new theorem provides an equally secure foundation for both versions of decision theory there will no longer be any reason to prefer one to the other on purely formal grounds. In the first instance, this should comfort and encourage the causal decision theorists because they no longer need to worry about the foundational deficiencies of the approach. Evidential decision theorists gain something too, though, for one of the main morals I wish to draw is that there is a deep sense in which Jeffrey's theory is exactly right: *all value is a kind of news value* even if not all kinds of news value are relevant to the choice of actions.

The plan of the book is as follows: Chapters 1–3 comprise a general introduction to expected utility theory that is meant to prepare readers for the discussion of evidential and causal decision theory that takes place in Chapters 4–7. Chapter 1 provides a quasi-historical introduction to expected utility theory as it applies to casino gambling, the case where it works best. It is meant primarily for readers who are coming to the subject for the first time. Those who already know a little bit about expected utility theory will miss nothing important by skipping the first chapter entirely. The second chapter clarifies the concept of a decision problem. My treatment here is slightly nonstandard since I follow Jeffrey in supposing that the components of decision problems are always *propositions* and yet still maintain a distinction among actions, states, and outcomes. Chapter 3 is an extended critical discussion of the influential formulation of expected utility theory that appears in Leonard Savage's *The Foundations of Statistics*. I shall argue that Savage's theory is not ultimately acceptable as a foundation for expected utility theory. This is an important conclusion in the present context because causal decision theorists have tended to assume that some appropriately modified version of Savage's theory would supply an adequate formal underpinning for their account. Since this is not the case, the need for a representation theorem for causal decision theory becomes all the more pressing.

Chapter 4 begins the heart of the book. In it I present Jeffrey's "evidentialist" version of decision theory and sketch Bolker's representation theorem for it. While Bolker's theorem will be seen to avoid nearly all the pitfalls that beset Savage's approach, it will be argued, nevertheless, that it is not entirely acceptable because it does not sufficiently constrain rational beliefs. A better version of Bolker's result will then be proved, one that, for the first time, obtains *unique* representations within Jeffrey's framework. This turns out to be a very important advance.

Chapter 5 treats the topic of causal decision theory. First I will argue that its proponents are correct in thinking that an adequate solution to Newcomb's problem requires an account of rational decision making that portrays agents as having beliefs about causal connections that are not ultimately reflected in their ordinary conditional subjective probabilities. The nature of these "causal" beliefs will be discussed at length. In the course of this investigation it will become apparent that all the various formulations of causal decision theory suffer from a common problem of "partition dependence"; they apply only when the decision situation is described in a very specific and detailed way. This makes it difficult to apply the theory to real-life decisions, and it greatly complicates the foundational challenges that it faces. The lesson will be that causal expected utility must assume the form of a conditional decision theory if the problem of partition dependence is to be solved.

The topic of Chapter 6 is the concept of *supposition* that underlies the notion of conditional expected utility. It is well known that there are at least two ways to suppose that a proposition is true: One can suppose it *indicatively* by provisionally adding the proposition to one's stock of knowledge, or one can suppose it *subjunctively* by imagining a possible circumstance in which the proposition is true that otherwise deviates minimally from the way that things actually are. I do not think the relationship between these two notions has yet been adequately understood, and the goal of Chapter 6 is to shed some light on the issue and, more generally, to clarify the concept of a supposition itself. In passing, I will show how the infamous "problem of old evidence" can be partially solved (or, better, how part of the problem can be completely solved), and a generalization of Jeffrey conditionalization to Rényi–Popper measures will be presented.

In Chapter 7 we examine a number of the representation results that have been proposed as foundations for causal decision theory. I shall argue that none of them is fully acceptable. I then go on to prove a representation theorem for conditional decision theory along the lines of Bolker's Theorem for evidential decision theory. This theorem will be seen to provide a common theoretical underpinning for both causal and evidential decision theories.

The book concludes with a short chapter that describes what has been accomplished and suggests some directions for future research.

By the way, I was kidding about the sucker gene. I decided to open the book with that example to weed out unsympathetic readers. It is good to know that you are still with me! Enjoy the fallacies!

1

Instrumental Rationality as Expected Utility Maximization

This chapter provides a brief quasi-historical introduction to *expected utility theory*, the most widely defended version of *normative decision theory*. The overarching goal of normative decision theory is to establish a general standard of rationality for the sort of *instrumental* (or "practical") reasoning that people employ when trying to choose means appropriate for achieving ends they desire. Expected utility theory champions *subjective expected utility maximization* as the hallmark of rationality in this means-ends sense.

We will examine the theory in the setting where it works best by applying it to the case of professional gamblers playing games of chance inside casinos. In this highly idealized situation, the end is always the maximization of one's own fortune, and the means is the ability to buy and sell wagers that offer monetary payoffs at known odds. Later chapters will consider more general contexts. Since the material here is presented in an elementary (and somewhat pedantic) way, those who already understand the concept of expected utility maximization and the rudiments of decision theory are encouraged to proceed directly to Chapter 2.

1.1 PASCAL AND THE "PROBLEM OF THE POINTS"

The *Port Royal Logic* of 1662 contains the first general statement of the central dogma of contemporary decision theory:

> In order to decide what we ought to do to obtain some good or avoid some harm, it is necessary to consider not only the good or harm in itself, but also the probability that it will or will not occur; and to view geometrically the proportion that all these things have when taken together.[1]

In modern terms, the suggestion here is that risky or uncertain prospects are best evaluated according to the *principle of mathematical expectation*, so that "our fear of some harm [or hope of some good]

---

[1] Arnauld and Nicole (1996, pp. 273–74).

ought to be proportional not only to the magnitude of the harm [or good], but also to [its] probability."[2]

This principle, and the theory of probability that underlies it, had been discovered in 1654 by Blaise Pascal, the greatest of the many great thinkers that Port Royal produced, during the course of a correspondence with Pierre de Fermat concerning a gambler's quandary now known as the *Problem of the Points*.[3] It had been posed to Pascal by a "reputed gamester," the Chevalier de Méré, who Pascal regarded as a fine fellow even though he suffered from the "great fault" of not being a mathematician. The question had to do with the fair division of a fixed pot of money among gamblers forced to abandon a winner-take-all game before anyone had won. Here is a simplified version of the problem that Pascal and Fermat considered (with dollars instead of "pistoles" as currency): Two gamblers, H and T, are playing a game in which a coin, known to be fair, is to be tossed five times and a $64 prize awarded to H or T depending on whether more heads or tails come up. Suppose that the first three tosses go head/tail/tail, and that the game is then interrupted, leaving the two gamblers with the task of finding an equitable way to dividing the $64. T, who has two of the three tails she needs to win, would surely feel cheated if the pot were split down the middle. H, on the other hand, would be justifiably upset if T got all the money since he still had a chance to win the game when it was stopped. Clearly, the fair division must give T something more than $32 but less than $64. The challenge for Pascal and Fermat was to find the right amount. Both men solved the instance of the Problem of the Points they were considering, but Pascal, in a great feat of mathematical genius, went on to treat the general case.

His solution had two parts. First, he invented the theory of probability more or less from scratch. Professional gamblers had long known that one could use nonnegative real numbers to measure the frequencies at which various events occur, and that these would give the odds at which various bets would be advantageous. Legend has it, for example, that the Chevalier de Méré made a lot of money laying even odds that he could roll at least one 6 in four tosses of a fair die. What the Chevalier realized, and his gullible opponents did not, was that the probability of this event was slightly more than one-half (about 0.518), and thus he was likely to win his bet more often than not. Unfortunately, probabilities were difficult to calculate, and gamblers were forced to find them empirically by observing the

---

[2] Arnauld and Nicole (1996, pp. 274–75).
[3] The best discussion of the Pascal/Fermat correspondence is found in Todhunter (1865/1949, pp. 7–21).

relative frequencies at which various events occurred. This method worked well for simple events, but it was hard to apply in even modestly complicated cases. In fact, after people had seen through his first scam, de Méré nearly bankrupted himself by overestimating his chances of throwing a 12 in twenty-four tosses of a pair of fair dice. Pascal solved the gamblers' problem by discovering

The Fundamental Law of Probability. *If $\{E_1, E_2, \ldots, E_n\}$ is any set of jointly exhaustive, mutually exclusive events, each of which has a definite probability, then the sum of all these probabilities is 1, that is,* $\Sigma_j \rho(E_j) = \rho(E_1) + \rho(E_2) + \ldots + \rho(E_n) = 1.$

While he never expressed it quite so explicitly, there is no doubt that Pascal understood this principle and recognized it as the key to calculating probabilities. He would also have endorsed the modern equivalent reformulation of his law in terms of the following three *axioms of probability*, which are supposed to hold for all events E and E^* (where $\neg$ and $\vee$ are the Boolean operations of negation and disjunction):

Nonnegativity. $\rho(E) \geq 0$
Normalization. $\rho(E \vee \neg E) = 1$
Finite Additivity. $\rho(E \vee E^*) = \rho(E) + \rho(E^*)$, whenever E and E^* are mutually incompatible events

He may also have endorsed the further axiom (which is of somewhat more recent vintage)

Continuity. For any denumerably infinite sequence of events E_1, E_2, $E_3, \ldots$ whose probabilities are well defined one has

$$\lim_{n \to \infty} \rho(E_1 \vee E_2 \vee \ldots \vee E_n) = \rho(E_1 \vee E_2 \vee E_3 \vee \ldots)$$

Note for future reference that Continuity and Finite Additivity ensure that probabilities are *countably additive* in the sense that $\Sigma_j \rho(E_j) = \rho(E_1 \vee E_2 \vee \ldots)$ whenever the E_j form a countable sequence of mutually incompatible events.

Pascal brought his new theory of probabilities to bear on games of chance by noting that any listing of the possible endings to a gambling game is always a *partition* of mutually exclusive, collectively exhaustive events. For the version of the Problem of the Points considered here, one partition of endgames (in obvious notation) is $\{E_{HH}, E_{HT}, E_{TH}, E_{TT}\}$, another is $\{(E_{HH}, (E_{HT} \vee E_{TH}), E_{TT})\}$, and a third is $\{E_{HH}, (E_{HT} \vee E_{TH} \vee E_{TT})\}$. Applied to these three cases, Pascal's basic insight was that

11

$$1 = \rho(E_{HH}) + \rho(E_{HT}) + \rho(E_{TH}) + \rho(E_{TT})$$
$$= \rho(E_{HH}) + \rho(E_{HT} \vee E_{TH}) + \rho(E_{TT})$$
$$= \rho(E_{HH}) + \rho(E_{HT} \vee E_{TH} \vee E_{TT})$$

Since the coin is fair $1/4 = \rho(E_{HH}) = \rho(E_{HT}) = \rho(E_{TH}) = \rho(E_{TT})$, and this determines the probabilities for all disjunctions involving these four events. For example, the probability that T had of winning the game had it been finished can be computed as $\rho(E_{HT} \vee E_{TH} \vee E_{TT}) = \rho(E_{HT}) + \rho(E_{TH}) + \rho(E_{TT}) = 3/4$.

Having thus invented probability theory, Pascal went on to propose that a player's fair share of the pot in a truncated game of points should always be her *expected payoff*, the quantity obtained by combining her winnings under all endgames "geometrically" (i.e., multiplicatively) with their probabilities. This means that a player who stood to win x_i if the game had ended in the way described by E_i would have a claim on $\rho(E_i)x_i$ from the pot, and her total fair share could be found by summing over all possible endgames to get her expected payoff $Exp_\$ = \$\Sigma_i\rho(E_i)x_i$. In our example, T, who wins if a tail is thrown on either of the final two tosses, is awarded

$$\$[\rho(E_{HH}) \times 0 + \rho(E_{HT}) \times 64 + \rho(E_{TH}) \times 64 + \rho(E_{TT}) \times 64] = \$48$$

while H, who needs two heads to win, is entitled to

$$\$[\rho(E_{HH}) \times 64 + \rho(E_{HT}) \times 0 + \rho(E_{TH}) \times 0 + \rho(E_{TT}) \times 0] = \$16$$

Pascal's solution to the Problem of the Points has many attractive features. First, since the expected payoffs for a group of players always add up to the amount in the pot, nothing ever goes to waste and no one is ever promised money that cannot be paid. Expected payoffs also have the right kind of symmetry properties for a "fair division" rule. It seems a minimum requirement of fairness that two players who had exactly the same chances of winning the same prizes when the game ended should be assigned the same share of the pot. The expected payoff scheme does this. Moreover, a division rule should never give one player a larger payoff than another if the second stood to win more money in *every* possible end-game. Indeed, if the second player were sure to win $y more than the first in every case, then it would seem that her payoff should be exactly $y larger. Again, expected payoffs have this desirable property. Finally, on Pascal's proposal, a player's fair share of the pot depends *only* on the monetary prizes she had a chance of winning when the game ended and the probabilities with which she might have won them.

This last point is particularly important because it entails that the only thing relevant to determining a bettor's share of the pot is the *wager* that she happens to hold when the game ends, where a "wager" is just a specification of her potential winnings and the circumstances under which she stands to win them. In our example, the wager T faced just before the game ended was

Event:	Head/Head	Head/Tail	Tail/Head	Tail/Tail
Probability:	1/4	1/4	1/4	1/4
Outcome:	T wins $0	T wins $64	T wins $64	T wins $64

Notice that this makes no reference to H's payoffs, to the overall size of the pot, or even to the history of the game. In any betting arrangement – no matter how many players are involved, no matter how much money is in the pot, no matter what kind of rules are in force – if T ends up in the situation described by the table then she is entitled to exactly $48 under Pascal's proposal.

One way to put this is to say that Pascal sets the *fair price* of the wager that T holds when the game ends at $48. This is the sum of money at which T should be equally happy to have either a straight payment of the sum or the wager itself. In general, a wager G's *fair price* for an agent is that amount of money $g at which she would be indifferent between holding G and having a straight payoff of $g. This price is characterized by the following two conditions:

(a) The agent finds the prospect of having any fortune greater than $g *strictly more desirable* than the prospect of holding G.
(b) She finds the prospect of having any fortune less than $g *strictly less desirable* than the prospect of holding G.

Economists sometimes refer to G's fair price as its "certainty equivalent" to indicate that it is a "riskless asset" that is equivalent in value to the "risky" prospect G.

Pascal's solution to the Problem of the Points thus has two parts: First, he claims that any bettor's fair share of the pot is the fair price of the wager she holds when the game ends. This makes sense given that the latter quantity is exactly what she could reasonably demand in compensation for ending the game voluntarily at that time. Next, he maintains that a wager's fair price should be identified with its expected payoff, the amount obtained by multiplying each of its prizes by its probability and then summing over the whole. The real substance of Pascal's proposal, then, is that *wagers should be valued according to their expected payoffs.*[4] To appreciate the implications of this idea we shall need to spend a moment getting clearer about the concept of a wager and the notion of an expected payoff.

1.2 WAGERS, PROBABILITIES, AND EXPECTATIONS

Pascal seems to have thought of wagers as arrangements under which a gambler is sure to end up having one of a finite set of (positive or negative) sums x_1, x_2, ..., x_n *added* to her net worth depending upon which member of a partition of mutually exclusive, jointly exhaustive events $E_1, E_2, \ldots, E_n$ happens to occur. Here is an example in which the event partition lists all of the possible ways in which a fair die might fall

Die Falls:	1	2	3	4	5	6
Incremental Payoff:	$11	−$6	−$13	−$9	−$5	$66

Under this arrangement, the bettor would have $11 added to her bank account if the die were to come up 1, $6 would be subtracted if it were to come up 2, $13 would be subtracted if were to come up 3, and so on.

We will conceive of wagers somewhat differently. First, rather than portraying payoffs as *changes* in wealth we will construe them as specifications of *levels of total wealth*. Thus, for a person whose current net worth is $10,000 the correct description of the above wager would be

Die Falls:	1	2	3	4	5	6
Fortune:	$10,011	$9,994	$9,987	$9,991	$9,995	$10,066

while for someone with a net worth of $25,000 it would be

Die Falls:	1	2	3	4	5	6
Fortune:	$25,011	$24,994	$24,987	$24,991	$24,995	$26,066

This may take some getting used to. Psychological research has shown that people typically assess economic prospects in terms of their *incremental* effects on wealth rather than their effects on total wealth.[5] We must be careful not to think this way, however, because we need to allow for a possibility that Pascal never envisioned, namely, that the attractiveness of a wager might depend on both the changes it can

---

[4] While the idea that wagers should be valued by their expected payoffs had occurred to Pascal as early as 1654, it was Christian Huygens who first defended it in print in his immensely influential 1659 textbook *De Ratiociniis in Ludo Aleoe.*

[5] For a discussion of the research in this area see Shafir and Tversky (1995).

14

make to a gambler's fortune and the size of the fortune she starts out with. Indeed, we shall shortly see that the idea of valuing wagers by their expected payoffs founders on precisely this point.

We will also go beyond Pascal by thinking of every wager as being defined over *the same* partition of events. Following the classic treatment in Leonard Savage's *Foundations of Statistics*, we imagine that there is a partition **S** of mutually exclusive, jointly exhaustive *states of the world* each of which specifies a determinate result for every contingency that could conceivably affect a bettor's fortune. So, for every coin ever tossed, every die ever thrown, every roulette wheel ever spun, every poker hand ever dealt, a state will say how the coin falls, how the die lands, where the wheel comes to rest, what cards are dealt, what cards a player would draw were he to draw them, and so on, and so on. While Savage left the notion of a state an unanalyzed primitive in his theory, we will follow the lead of Richard Jeffrey and construe states as *propositions*. I will say more about what this means in the next chapter, but it suffices for now to think of state propositions as *descriptions* of possible states of affairs. These descriptions can properly be called true or false, can be combined using Boolean connectives and other propositional operations, and can be used to represent the ultimate objects of the bettor's beliefs. Every state proposition will describe a possible course of events that is *complete* in the sense that it specifies a unique, determinate result for every wager that anyone might have occasion to consider. Thus, the states in **S** should be so finely specified that a bettor's uncertainty about the outcomes of the wagers she holds can always be modeled as uncertainty about which state actually obtains.

Events are less specific propositions about possible world histories that can be expressed as disjunctions of states. For technical convenience, we will suppose that the set of all events or *event space*, here denoted $\Omega(\mathbf{S})$, is a σ-*complete Boolean algebra*. This is a set of propositions that is closed under negation and countable disjunction, so that $\neg E_j$ and $\vee_j E_j = (E_1 \vee E_2 \vee E_3 \vee \ldots)$ are events whenever E_1, E_2, $E_3 \ldots$ are events.

A wager, for current purposes, is simply a proposition that describes a way in which a bettor's total fortune depends on the state of the world. Such a proposition can be expressed as a conjunction of subjunctive conditionals of the form[6]

---

[6] Notice that the expression "$\$g$" is doing double duty here. It signifies both the sum of money *g dollars*, and the proposition *that the agent's total fortune is g dollars*. This ambiguity should cause no confusion since the context will always determine which of the two is meant.

$$G = \&_s \left(\begin{array}{l} \text{If } S \text{ were true, then the gambler's} \\ \text{total fortune would be } \$g_s \end{array} \right)$$

$$= \&_s (S \,\square\!\!\rightarrow \$g_s)$$

where S ranges over states of the world, and "$\square\!\!\rightarrow$" stands for the subjunctive conditional "If__were the case, then ---- would also be the case."

Economists make a distinction between two kinds of wagers or, better, between two states of a gambler's knowledge about the events on which they depend. When a wager involves *risk* each state S has a definite objective probability $\rho(S)$ that is known to the bettor. In contrast, *uncertain* prospects are those in which the bettor is in the dark about the objective probabilities of at least some of the events on which her fortunes depend, either because she does not have access to them or because there are no such probabilities to be known. The paradigm example of an uncertain prospect is a bet on the finish of a horse race (described here for an agent with an initial fortune of $10,000)

Events:	Stewball wins.	Stewball places.	Stewball shows.	Stewball runs in the field.
Payoffs:	$10,012	$10,006	$10,003	$10,002

Unlike the other cases we have considered, here it is unrealistic to expect an agent to know the probabilities of the events that will determine her fortune since the factors that decide horse races are too numerous and unpredictable to make knowledge of objective chances possible (even if they do exist).

Pascal was concerned, at least initially, with finding fair prices for wagers involving risk rather than uncertainty. When considering the Problem of the Points, for example, he assumed that each endgame would have a definite probability that would be known to all players. We will see that Pascal did consider extending his approach to broader contexts, but for now it will be useful for us to think of him as offering a theory of fair pricing for wagers involving risk. We shall thus assume that there is a function ρ that assigns each event E in the algebra $\Omega(\mathbf{S})$ an *objective probability* $\rho(E)$ between 0 and 1, and that our gamblers all know exactly what these probabilities are.

Once we have an objective probability function in hand we can define *objective expected values* for many functions that assign real numbers to states in $\mathbf{S}$. A real-valued function F of the states in $\mathbf{S}$ is said to be *measurable* with respect to ρ if and only if the disjunctive

16

proposition $F_x = \vee\{S \in \boldsymbol{S}: F(S) = x\}$ is in ρ's domain $\Omega(\boldsymbol{S})$ for each real x. In other words, F is measurable exactly if ρ assigns a definite probability to every proposition of the form $F_x = $ "F's value is x." When F is measurable it makes mathematical sense to define its ρ-*expectation* as $\mathrm{Exp}_\rho (F) = \int_S F(S)d\rho$. This integral is nothing more than a weighted average of F's values where the weight assigned to each value is the probability of the state that generates it.[7] In the cases that will be of most interest to us, F's *support*, the set of values whose associated F_x propositions have nonzero probability, will be at most countably infinite and the preceding expression will reduce to a straight sum of those values of F that have a nonzero probability of being realized weighted by their probabilities, so that $\mathrm{Exp}_\rho(F) = \Sigma_x\, x\rho(F_x)$ where x ranges over the values in F's support. Except in rare cases, which will be clearly noted, the reader will never go wrong thinking of expectations as sums of this sort. In what follows I will largely ignore the difference between integrals and sums, and will write $\mathrm{Exp}_\rho(F) = \Sigma_S F(S)\rho(S)$. Strictly speaking, this identity will hold only when there are countably many states, but everything I will have to say about such sums will carry over to integrals of the form $\int_S F(S)d\rho$.

For future reference, note that expectations obey the following principles:

Constancy. If $F(S) = 1$ for all states S, then $\mathrm{Exp}_\rho(F) = 1$.

Dominance. If $F(S) \geq F^*(S)$ for all states S, then $\mathrm{Exp}_\rho(F) \geq \mathrm{Exp}_\rho(F^*)$, and this inequality is strict when there is a nonzero probability that F's value will be larger than F*'s.

Additivity. $\mathrm{Exp}_\rho(F + F^*) = \mathrm{Exp}_\rho(F) + \mathrm{Exp}_\rho(F^*)$ provided that the latter two expectations are defined.

Continuity. If $\mathrm{Exp}_\rho(F_j)$ is defined for each $j = 1, 2, 3, \ldots$, and if the sequence of functions $F_1, F_2, F_3, \ldots$ converges to F, then $\lim_{j\to\infty}\mathrm{Exp}_\rho(F_j)$ converges and is equal to $\mathrm{Exp}_\rho(F)$.

It is straightforward to show that these imply that the expectation operator is *linear*, so that $\mathrm{Exp}_\rho(cF + c^*F^*) = c\mathrm{Exp}_\rho(F) + c^*\mathrm{Exp}_\rho(F^*)$ for any real numbers c and c^*. They also entail that expectations are *countably additive* in the sense that $\mathrm{Exp}_\rho(\Sigma_j F_j) = \Sigma_j\mathrm{Exp}_\rho(F_j)$, provided that all estimates are well defined and the right-hand sum converges.

There is a tight connection between wagers and real functions defined over states of the world. Each wager $G = \&_S (S\ \square\!\!\rightarrow\ \$g_S)$ uniquely picks out a function $G(S) = g_S$, and any such function G seems to specify an associated wager $G = \&_S(S\ \square\!\!\rightarrow\ \$G(S))$. Indeed,

---

[7] For readers interested in the details, I recommend Royden (1968).

17

the connection is so close that many decision theorists *identify* wagers with abstract state-to-outcome functions.[8] I prefer not to do this both because I want to insist on a conceptual distinction between the propositions that describe these mappings and the mappings them- selves, and because I shall eventually reject the notion that every mapping from states to outcomes has a well-defined wager associated with it. The converse, however, is true, and we can use it to define a wager's expected payoff as follows:

Definition. *If $G = \&_S (S \square\rightarrow \$g_S)$ is a wager and $G(S)$ is its associated state-to-payoff function, then G's expected payoff, or actuarial value, relative to the probability function ρ is $Exp_\rho(G) = \Sigma_{S \in S} G(S)\rho(S)$ (or $\int_S G(S)d\rho$ when the set of states is uncountable).*

With this definition in hand we are in a position to give precise statement of Pascal's fair pricing scheme for risky wagers.

Pascal's Thesis. *If a gambler believes that ρ is the correct objective probability function for states of the world, and if she is interested in the wager G solely as a means of increasing her total fortune, then she should find the prospect of holding G and the prospect of having a sure fortune equal to G's objective expected payoff equally desirable, so that for her $\$g = Exp_\rho(G)$.*

I have included the proviso that the bettor should be interested in G "solely as a means of increasing her fortune" because, as I think Pascal recognized, his thesis is plausible only when applied to "professional" gamblers who value nothing but money (at least when inside the casino), and who derive neither pleasure nor displeasure from the act of betting itself. Such a person will care about a wager only insofar as its truth might affect the size of her fortune. One way to put this is to say that she will regard money as *an unalloyed good* in the sense that she will find the prospects ($\$x \& E$) and ($\$x \& \neg E$) equally desirable for any event E and any wealth level $\$x$.

There are no "professional" bettors of course; people value money for what it can buy, not as an end in itself. This would be a serious problem if we were interested in applying Pascal's Thesis to real human beings, but this is not the plan. Our aim, instead, will be to see what the thesis tells us about the highly idealized, and thus more tractable, case of professional gamblers in the hopes of discovering

---

[8] The classic reference here is Savage (1954/1972).

general principles of instrumental rationality that can be applied in the more realistic settings to be considered in later chapters.

When combined with the basic properties of expectations, Pascal's Thesis imposes a number of restrictions on a rational professional bettor's fair prices.

Constancy.   A wager $G = \&_s(S \,\square\!\!\rightarrow \$c)$ that is sure to leave a bettor with a fortune of $\$c$ in every state of the world has $\$c$ as its fair price.

Dominance. If $G = \&_s(S \,\square\!\!\rightarrow \$g_s)$ is sure to award at least as large a fortune as $H = \&_s(S \,\square\!\!\rightarrow \$h_s)$ does, so that $g_s \geq h_s$ for all $S \in \mathbf{S}$, then G's fair price will be at least as great as H's. Moreover, if there is a positive probability that G will produce a fortune strictly larger than H's, then G's fair price will be strictly higher than H's.

Additivity.   If $G = \&_s(S \,\square\!\!\rightarrow \$g_s)$ and $G^* = \&_s(S \,\square\!\!\rightarrow \$g_s^*)$ have fair prices of $\$g$ and $\$g^*$, respectively, then the fair price of the wager $H = \&_s(S \,\square\!\!\rightarrow \$(g_s + g_s^*))$ is $\$h = \$(g + g^*)$.

Continuity. Let $g_s(1)$, $g_s(2)$, $g_s(3)$, ... be real numbers converging to g_s for each $S \in \mathbf{S}$. Then, the fair price of $G = \&_s(S \,\square\!\!\rightarrow \$g_s)$ is the limit of the fair prices of the wagers $G(j) = \&_s(S \,\square\!\!\rightarrow \$g_s(j))$.

We will be evaluating each of these principles. But before doing so we need to correct a misconception about the nature of fair prices that has caused a great deal of confusion about the status of requirements of the sort just presented.

1.3 A SHORT DIGRESSION: AGAINST BEHAVIORISM

I was careful to formulate Pascal's Thesis as *a norm of rational desire* that governs the fair pricing of risky wagers, where fair prices are understood in terms of the pattern of preferences described in (a)–(b) of Section 1.1. I framed the issue this way because I hoped to distinguish my version of Pascal's Thesis from a closely related principle that governs not desires, but *overt actions*. In the economics literature one often finds fair prices defined not in terms of desires, but in terms of *overt choice behavior*: Instead of (a)–(b), the conditions for a gambler's having $\$g$ as her fair price for G become

(a*)   If asked to *choose* between holding G and having a guaranteed fortune of any sum greater than $\$g$, the gambler would choose the guaranteed fortune (provided that she is certain that nothing else of value hangs on her choice).

19

(b*)   If asked to *choose* between holding *G* and having a guaranteed fortune of any sum less than $*g*, the gambler would choose the wager (provided that she is certain that nothing else of value hangs on her choice).

This leads to the following revision:

Pascal's Thesis (as a constraint on acts). *If a gambler believes that ρ is the correct objective probability function for states of the world, and if she is interested in a wager G only as a means of increasing her fortune, then she should be willing to exchange G to attain any wealth level that is not less than $Exp_\rho(G)$, and she should be willing to exchange any wealth level that is not greater than $Exp_\rho(G)$ to obtain G.*

The two versions of Pascal's Thesis are closely related – anyone who thinks that gamblers should obey the first will agree that, *insofar as they are going to be in the business of buying and selling wagers*, they should obey the second. Nevertheless the two principles are conceptually distinct and the notion of a fair price defined in (a)–(b) is not the same as that defined in (a*)–(b*). In saying this I am departing from what has been the party line among rational choice theorists for many years. Decision theory came of age during the heyday of logical positivism, and many of its early practitioners embraced a kind of behaviorism that equated desires with *dispositions toward overt action*. They thus took it as *analytic* that an agent would have desires satisfying (a)–(b) if and only if she also had the behavioral dispositions described in (a*)–(b*).

It is easy to see why such a view might appeal to psychologists, economists, market researchers, or bookies interested in making an empirical study of the prices people pay for wagers. The challenge for these investigators is to find ways of making reliable inferences about the "internal" states and processes implicated in decision making on the basis of empirical information about patterns of overt behavior. Behaviorists solve this problem by adapting logical positivism's general approach to knowledge about unobservables to the case of psychological states. The positivists held that inferences from observable facts about, say, falling bodies to unobservable facts about the gravitational field must be made via the use of analytically true "bridge laws," "coordinative definitions," or "meaning postulates" that give (at least part of) the meaning of "gravitational field" by listing the empirical consequences of various sentences in which the term appears. Behaviorism proposes something similar for desires (and beliefs). It is supposed to be *analytic* that any agent who desires one prospect more than another will be disposed to act in ways that are likely to produce the first rather than the second in any context where only one of the two can come about. This makes it a *misuse of language* to say that a person whose fair price for G is $g violates (a*)–(b*). The advantage of this rather radical stance is that it allows investigators to establish secure conclusions about a person's fair prices by watching what she does. When we see her giving up a sure fortune of $x to buy a wager in a situation where nothing else of value hangs in the balance, we can be certain that her fair price for the wager is greater than $x, or so behaviorism has it.

Despite its attractiveness, the behaviorist interpretation of rational choice theory is slowly becoming a thing of the past (and should be made entirely a thing of the past as soon as possible). The old guard still insists that the concept of a fair price can only be understood in terms of behavioral dispositions, but it has become clear that the theoretical costs of this position far outweigh its benefits. There are just too many things worth saying that cannot be said within the confines of strict behaviorism. One cannot, for example, say whether an agent is disposed to satisfy (a*)–(b*) because she regards G and $g as equally desirable for their own sakes or because she believes that those who treat the less desirable G on a par with the more desirable $g will be rewarded in the hereafter. The basic difficulty here is that it is impossible to distinguish contexts in which an agent's behavior really does reveal what she wants from contexts in which it does not without appealing to additional facts about her mental state. Indeed, the only cases in which (a*)–(b*) are sure to indicate that a person finds G and $g equally desirable are those in which she is certain either that her choice will generate no consequences other than G or $g or that any consequences generated will have an equal effect on the desirabilities of these two propositions. To say this, however, is to go well beyond what the behaviorist interpretation allows since we are now imputing mental states to the agent that will not ultimately be manifested in her overt choice behavior.

An even more serious shortcoming is behaviorism's inability to make any sense of *rationalizing explanations* of choice behavior. When an agent is disposed to buy G for $x it always makes sense to inquire into her reasons. Does she have a *rationale* for making the purchase? If so, what is it? The answer, of course, is usually going to be that she believes this transaction will result in her gaining G and losing $x, *and she finds this state of affairs preferable to one in which she does not gain G and keeps $x.* If this explanation is to avoid circularity it must be possible to understand the italicized phrase as

21

meaning something other than that she is willing to exchange G for x. So, insofar as we want to be in the business of giving rationalizing explanations of behavior, we must recognize that there is a conceptual difference between finding G and g equally desirable and being disposed to buy or sell G for g. More generally, since we shall often want to use (a)–(b) as a part of a noncircular rationalizing explanation for (a*)–(b*) we cannot simply regard the two as different ways of saying the same thing.

None of this is meant to deny that there is a tight connection between fair prices and betting behavior. An agent who has g as her fair price for G will often be willing to buy the wager at prices not greater than g and sell it at prices not less than g. Moreover, under the right conditions one *can* use this fact to draw reliable inferences about her fair prices on the basis of her behavior. However, unlike the behaviorist, who treated such inferences as analytic, we must recognize them as *inductions* from facts about choices to distinct facts about what the agent finds desirable. A person's betting behavior is often a highly reliable indicator of her fair prices, but it does not constitutively determine them.

The moral here is that rational choice theory is not so much about overt choices as about the underlying *desires and beliefs* that cause and rationalize them. When we criticize or commend any action on grounds of practical rationality we are not assessing the actor's behavior per se, but are evaluating the pattern of desires and beliefs that brought it about. More precisely, when we say that it was irrational, in the instrumental sense, for an agent to perform A we mean that she had an all-things-considered desire to do A, this desire was her motive for doing A, and it was irrational for her to want A given her other desires and beliefs. So, whenever some decision theorist says that a given action is irrational what he or she must really mean is that the agent's *all-things-considered desire to perform it* is irrational (in the sense that it does not cohere with her other beliefs and desires). It rarely does harm to speak of acts as rational and irrational without mentioning the underlying desires, but this is only because people so often do what they want. We must keep in mind that the connection between wanting and doing is not analytic. An inference from "The agent has an all-things-considered desire to perform A" to "She would perform A if given the chance" can be very strong in the inductive sense, as can its converse, but neither is grounded in any kind of conceptual necessity. For these reasons, our official version of Pascal's Thesis will be the one expressed in terms of desire rather than action.

Pascal never offered any systematic rationale for the policy of using expected payoffs as fair prices. This is not surprising. Proponents of new scientific ideas invest almost all their time in problem-solving activities; foundational issues only come to the fore after all the low-hanging fruit has been picked. With the hindsight of those fortunate enough to live after Frank Ramsey, Bruno de Finetti, Leonard Savage, John von Neumann, and Otto Morgenstern, it is now possible to see how a justification for Pascal's Thesis might go.[9] We cannot, of course, be sure that it is the sort of justification Pascal would have offered had he thought seriously about the matter. (In fact, it is likely that he would have given up on the idea that fair prices are expected payoffs once he fully understood what it entailed.) Be this as it may, the following argument does present the best case that can be made for Pascal's Thesis. It is crucial that the reader have a clear sense of how it goes because a large part of this book will be spent improvising on its basic theme. I will present the argument rather pedantically, highlighting every relevant premise, so as to render the overall structure of the reasoning as transparent as possible.

Let's begin by saying something about the set $\mathbf{G}$ of wagers over which fair prices are to be defined. The argument requires $\mathbf{G}$ to be a *mixture space*. An *even* or *(1/2,1/2) mixture* of two wagers G and G^* is a third wager H that always pays out a fortune of $\$x$ with probability $\frac{1}{2}\rho_1 + \frac{1}{2}\rho_2$ when G_1 pays $\$x$ with probability ρ_1 and G_2 pays $\$x$ with probability ρ_2. To illustrate, suppose that a nickel and a dime are going to be tossed, and consider

Nickel/Dime $\rho=$	Head/Head 1/4	Head/Tail 1/4	Tail/Head 1/4	Tail/Tail 1/4
G_1	$3	$2	$3	$2
G_2	$5	$6	$5	$6
H_1	$3	$2	$5	$6
H_2	$6	$3	$2	$5

Here *both* H_1 and H_2 are even mixtures of G_1 and G_2 because each yields G_1's payoffs with 1/2 probability and G_2's payoffs with 1/2 probability. One can often think of such a mixture as a compound wager that offers G_1 and G_2 as "prizes" with equal probability. The third line of the table suggests such an interpretation: H_1 pays G_1 if the

---

[9] See Ramsey (1931), de Finetti (1964), von Neumann and Morgenstern (1953), and Savage (1954/1972).

nickel falls heads and G_2 if it falls tails. This way of looking at things has limits, however, since there are mixtures, like H_2, that cannot be understood as compounds.[10]

We can generalize the notion of an even mixture by imagining that the nickel in our example is not fair. If its bias toward heads is $\lambda/(1 - \lambda)$ where $1 \geq \lambda \geq 0$ (so that the ρ row of the table is: $\lambda/2, \lambda/2, (1 - \lambda)/2, (1 - \lambda)/2)$, then H_1, but not H_2, is a $(\lambda, 1 - \lambda)$ mixture of G_1 and G_2. Here is the formal definition:

> H is a $(\lambda, 1 - \lambda)$ *mixture* of G_1 and G_2 if and only if, for every real number $x > 0$, H pays \$$x$ with probability $\lambda\rho_1 + (1 - \lambda)\rho_2$ whenever G_1 pays \$$x$ with probability ρ_1 and G_2 pays x with probability ρ_2.

Mixtures have an illuminating geometrical interpretation. If we think of G_1 and G_2 as "points" in the "space" of all wagers **G**, then a $(\lambda, 1 - \lambda)$ mixture of G_1 and G_2 lies on the "line segment" connecting them. An even mixture lies at the midpoint of the segment, a (3/4, 1/4) mixture lies halfway between G_1 and the midpoint, a (1/4, 3/4) mixture lies halfway between G_2 and the midpoint, and so on.

We can now state our first premise.

Premise 1 (Richness). *(i) For any sum of money \$x, **G** contains a wager that pays \$x in every possible circumstance; (ii) **G** is closed under mixing in the sense that for any $G_1, G_2 \in$ **G** and $\lambda \in [0,1]$ there is $H \in$ **G** such that, for every real number x, H pays \$x with probability $\lambda\rho_1 + (1 - \lambda)\rho_2$ when G_1 pays \$x with probability ρ_1 and G_2 pays \$x with probability ρ_2.*

This says that the set of wagers is a *mixture space*: It contains all straight payoffs (think "constant wagers"), and it contains any mixture of any wagers it contains.

Pascal's Thesis presupposes that each wager in **G** will have a *single* fair price that holds good for all agents. There are two independent claims here.

Premise 2 (Existence). *Each rational professional gambler will have a unique fair price for every wager in **G**; that is, for each $G \in$ **G** there*

---

[10] The mixture-as-compound view also has the unfortunate effect of leading people to think that mixtures have a temporal structure. One tends to look at H_1 and think, "First the nickel is tossed to decide whether the gambler gets G_1 or G_2; then the dime is tossed to determine her fortune." For a professional bettor the sequence of the events that instantiate a given wager is immaterial to its desirability except insofar as it determines the probabilities with which she will receive various fortunes.

must be a sum of money $g such that she finds holding G and having a fortune of $g equally desirable.

Premise 3 (Invariance). *Each rational professional gambler must have* the same *fair price for* every *wager in* **G***.*

Existence requires bettors to have highly *determinate* desires, so determinate, in fact, that they can put a precise price on every wager they consider. This leaves no room at all for vagueness or imprecision in judgments of value; for every sum of money $x there must be a fact of the matter about whether the bettor prefers G to $x, prefers $x to G, or finds the two prospects equally desirable. Invariance forces every bettor to value money in the same way. One way to put this is to say that a fixed sum of money should "buy the same amount of happiness" for everyone.[11] Neither of these requirements is very plausible, and we shall reject both of them later.

The next premise is much more compelling. It simply says that rational professional bettors place a higher value on wagers that are sure to pay them more money.

Premise 4 (Dominance). *If one wager $G = \&_s(S \ \Box \rightarrow \$g_s)$ dominates* another $H = \&_s(S \ \Box \rightarrow \$h_s)$ *in the sense that* $g_s \geq h_s$ *for all states S, then G's fair price is at least as great as H's. If, in addition,* $g_s > h_s$ *for all states S that entail some event E with positive probability, then G's fair price is strictly greater than H's.*

Assuming that our gambler always prefers more money to less, this just says that she should never put a higher price on G^* than on G unless there is some chance that G^* will leave her richer than G will. For future reference note that Dominance entails that the fair price of every wager must fall somewhere between its highest and lowest payoffs.

By way of introduction to the next premise consider the wagers

Nickel/Dime	Head/Head	Head/Tail	Tail/Head	Tail/Tail
ρ	3/8	1/8	3/8	1/8
G	$30	$20	$10	$30
G^*	$10	$30	$30	$20

[11] In speaking this way we must not be misled into thinking that the ability to buy happiness is what *makes* having money desirable for professional gamblers. The order is reversed. Since a professional gambler takes money to be a *basic* good, it buys her happiness because she desires it, *not* the other way around. This is what makes "professional" gamblers so unlike you and me.

where the dime is biased $3:1$ in favor of heads. Notice that G and G^* offer the same prizes with identical probabilities. As a consequence, someone who does not take them to be equally desirable must be basing her preference on a distinction between the two wagers that cannot be traced to differences in their payoffs or probabilities. She might, for example, prefer having a fortune of \$30 when the coins fall the same way to having that fortune when one comes up heads and the other comes up tails. Differences of this kind should matter only to a person whose interests in events outrun her interests in the fortunes they might produce. This never happens with professional gamblers. They treat money as an unalloyed good, so that for any event E and any wealth level \$x they find the prospect (\$x & E) exactly as desirable as the prospect (\$x & $\neg E$). Our next premise codifies this point.

Premise 5 (Stochastic Equivalence). *If G and H are stochastically equivalent in the sense that, for every real number x, G pays \$x with probability ρ if and only if H also pays \$x with probability ρ, then G and H have the same fair price.*

The effect of this is to rule out differences in desirability among wagers that differ only in the identities of the chance events that underlie their probability assignments.

This makes it possible for us to think of fair prices as attaching not to individual elements of $\boldsymbol{G}$ but to classes of stochastically equivalent wagers, where each such class may be identified with the probability distribution that is common to all its wagers. The main advantage of this is that it lets us pretend that a single object, $\lambda G_1 \oplus (1 - \lambda)G_2$, is *the* $(\lambda, 1 - \lambda)$-mixture of G_1 and G_2. $\lambda G_1 \oplus (1 - \lambda)G_2$ is thus a representative for any wager of the form

	E	$\neg E$
	λ	$1 - \lambda$
$\lambda G_1 \oplus (1 - \lambda)G_2$	G_1	G_2

What $\lambda G_1 \oplus (1 - \lambda)G_2$ really is, of course, is the probability distribution common to every wager H that can be expressed as a $(\lambda, 1 - \lambda)$-mixture of any wagers stochastically equivalent to G_1 and G_2. But, since the identities of the events in H do not matter except insofar as they have the probabilities λ and $1 - \lambda$, it makes sense to speak about these probabilities themselves, and thus to treat $\lambda G_1 \oplus (1 - \lambda)G_2$ as a single wager.

The following three principles express the basic formal properties of mixtures thought of in this abstract, event-independent way:

Mix₁. $G_1 = 1G_1 \oplus 0G_2$
Mix₂. $\lambda G_1 \oplus (1 - \lambda)G_2 = (1 - \lambda)G_2 \oplus \lambda G_1$
Mix₃. If $H = (\mu G_1 \oplus (1 - \mu)G_2)$, then $(\lambda H \oplus (1 - \lambda)G_2) = (\lambda\mu G_1 \oplus (1 - \lambda\mu)G_2)$.

All three are easily proven consequences of Stochastic Equivalence.

Our next premise is the linchpin of the entire theory, for it is what ensures that fair prices are governed by the laws of mathematical expectation. To motivate it, let's look at the following five wagers (where reference to the events have been suppressed because they do not matter):

ρ	3/8	1/8	3/8	1/8
G	$100	$40	$100	$40
G_1	$30	$20	$30	$20
G_2	$10	$50	$10	$50
H_1	$30	$20	$100	$40
H_2	$10	$50	$100	$40

Observe that $H_1 = \frac{1}{2}G_1 \oplus \frac{1}{2}G$ and $H_2 = \frac{1}{2}G_2 \oplus \frac{1}{2}G$. Since both wagers offer G as a "prize" with probability 1/2 it is natural to think that this common component should "drop out," so that any difference in desirability between H_1 and H_2 would be traceable to differences in the desirabilities of G_1 and G_2. This is what our next premise tells us.

Premise 6 (Independence). *For $\lambda > 0$, the fair price of $\lambda G_1 \oplus (1 - \lambda)G$ exceeds (is equal to) the fair price of $\lambda G_2 \oplus (1 - \lambda)G$ if and only if the fair price of G_1 exceeds (is equal to) the fair price of G_2.*

This imposes a kind of noncontextuality constraint on rational desires. Since the identity of G is immaterial, the principle says that the relative strengths of a bettor's desires for G_1 and G_2 should not vary when they are mixed in the same proportion with any third wager. This makes it irrational to give more weight to the desirability of G_1 than to that of G_2 when the two appear in $H_1 = \lambda G_1 \oplus (1 - \lambda)G$ and $H_2 = \lambda G_2 \oplus (1 - \lambda)G$ but to weigh G_2 more heavily than G_1 when they appear in, say, $H_1^* = \lambda G_1 \oplus (1 - \lambda)G^*$ and $H_2^* = \lambda G_2 \oplus (1 - \lambda)G^*$. We will discuss some of the reasons for accepting this principle in the next chapter.

27

In the presence of the other premises Independence entails the following important consequences (whose proofs are straightforward and will be left to the reader[12]):

Lemma 1.1a (Substitution). *$H_1 = \lambda G \oplus (1 - \lambda)G^*$ and $H_2 = \lambda \$g \oplus (1 - \lambda)G^*$ have the same fair price when $\$g$ is G's fair price.*

Lemma 1.1b (Stochastic Dominance). *If G's fair price is greater than G^*'s, then the fair price of $H = \lambda G \oplus (1 - \lambda)G^*$ increases monotonically with λ.*

Lemma 1.1c (Averaging). *If $\lambda, \mu, \nu \in [0,1]$ and if*
- *$\$g_\lambda$ is the fair price of $G_\lambda = \lambda\$x \oplus (1 - \lambda)\$y$*
- *$\$g_\mu$ is the fair price of $G_\mu = \mu\,\$x \oplus (1 - \mu)\$y$*
- *$\$h$ is the fair price of $H = \nu\$g_\lambda \oplus (1 - \nu)\$g_\mu$*

then $\$h$ is also the fair price of $H^* = [\sigma\$x \oplus (1 - \sigma)\$y]$ where $\sigma = (\nu\lambda + (1 - \nu)\mu)$.

According to Lemma 1.1a, it should never matter to a rational gambler whether she stands to get G with a certain probability or its fair price with that same probability. 1.1b is a formal way of saying that prospects offering higher probabilities of better outcomes are more desirable. While 1.1c looks complicated, all it really says is that the fair price of a $\nu/(1 - \nu)$ mixture of the fair prices of G_λ and G_μ is also the fair price of a $\nu/(1 - \nu)$ mixture of G_λ and G_μ themselves.

As a consequence of Lemma 1.1b, the fair price of $G_\rho = \rho\$x \oplus (1 - \rho)\$y$ increases monotonically in ρ when $x > y$. Our next premise ensures that this increase is a continuous one, so that small changes in probability never lead to large changes in G_ρ's fair price.

Premise 7 (Continuity). *Given any real numbers $x_1 > y > x_0$, there is a probability ρy such that $\$$ is the fair price of $\rho_y \$x_1 \oplus (1 - \rho_y)\$x_0$.*

Lemma 1.1b ensures that ρy is unique.

We are now in a position to state what is perhaps the single most important result in expected utility theory:

Theorem 1.2. *Given any real scaling constants $u > z$, if Premises 1–7 hold then there exists a unique monotonically increasing real function u such that*

---

[12] Substitution follows from the concept of a fair price via two applications of Independence. For the rest one may consult Savage (1954/1972, pp. 70–73) or Kreps (1988, pp. 43–51).

- $u(u) = 1$ and $u(z) = 0$
- The fair price $\$g$ for any wager $G = \&_s (S \square \rightarrow \$g_s)$ is the unique solution to the equation $u(g) = U(G)$ where $U(G)$ is the *expected value* of $u(g_s)$ computed relative to ρ, so that $U(G) = \int_S u(g_s)d\rho$, or, when the set of states is countable, $U(G) = \Sigma_s u(g_s)\rho(S)$.

The proof of this result is too technical to be conveniently included in an introductory chapter of this sort. Readers interested in the details may consult any one of a number of excellent sources,[13] or they might just read on since subsequent chapters will present proofs of more general theorems that entail this one as a special case.

The map u is not something Pascal would have recognized. It is called a *utility* function for money and should be interpreted as a measure of the overall "amount" of happiness or satisfaction that various sums of money would be able to buy (for a professional bettor). Theorem 1.2 shows that, subject to the arbitrary choice of a zero point z and unit u to set the scale for measuring utility, Premises 1–7 guarantee the existence of a unique function u that makes the expected utility of any wager coincide with the utility of its fair price. This means that a bettor who satisfies the premises will set fair prices as if she is calculating each wager G's expected utility and then aiming to find a sum of money $\$g$ whose utility is the same as $U(G)$. This is not to say that she actually employs this highly abstract process when she goes about fixing fair prices. The point is only that from the *third person perspective* it will look *as if* her fair prices have been arrived at in this way.

One illuminating way to understand Theorem 1.2 is by recognizing that the function u provides an *interval scale* on which to measure the "value" of money.[14] To see what this involves, let G be a wager that offers fortunes of $\$x$ and $\$y \neq \$x$ with probabilities ρ and $1 - \rho$ respectively. For two utility functions u and u* to generate the same

---

[13] For example: Fishburn (1970), Savage (1954/1972), Kreps (1988).
[14] The paradigmatic examples of interval scales are the Celsius and Fahrenheit measures of temperature. Each scale assigns "degrees" of temperature to physical objects in such a way that the thermal ordering among objects is invariant. While ratios of individual temperatures are not meaningful since they vary with the choice of scale, ratios of temperature differences are cardinally significant. So, while it makes sense to claim that an object at 100°F (= 37.38°C) is *hotter than* one at 50°F (= 10°C), it does not make sense to say that the first is twice (or 3.738 times) as hot as the second. One can, however, say that the *temperature difference* between a body at 50°F and one at 100°F is twice as great as the temperature difference between a body at 25°F (= −3.89°C) and one at 50°F because the ratio [100°F − 50°F]/[50°F − 25°F] = [37.38°C − 10°C]/[10°C − (−3.89°C)] = 2 is invariant among all legitimate ways of measuring temperature.

fair price for G there would need to be a real number $x \leq g \leq y$, such that $u(\$g) = \rho u(\$x) + (1 - \rho)u(\$y)$ and $u^*(\$g) = \rho u^*(\$x) + (1 - \rho)u^*(\$y)$. This can happen only if

$$\frac{u(\$g) - u(\$y)}{u(\$x) - u(\$y)} = \rho = \frac{u^*(\$g) - u^*(\$y)}{u^*(\$x) - u^*(\$y)}$$

In other words, the *ratio* of the *difference* in utility between $\$g$ and $\$y$ to the *difference* in utility between $\$x$ and $\$y$ must be an *invariant* quantity among all functions that satisfy $u(g) = U(G)$. The reader is invited to use this observation to verify that the following three conditions are equivalent.

(A) For all wagers $G \in$ **G** and all real numbers x, $u(x) \geq U(G)$ if and only if $u^*(x) \geq U^*(G)$.

(B) Ratios of expected utilities are invariant between u and u*, that is, for any wagers G_1, G_2, G_3, and G_4 with $\$g_3 \neq \$g_4$ one has

$$\frac{U(G_1) - U(G_2)}{U(G_3) - U(G_4)} = \frac{U^*(G_1) - U^*(G_2)}{U^*(G_3) - U^*(G_4)}$$

(C) u and u* are positive linear transformations of one another, so that $u^*(x) = mu(x) + b$ where $m > 0$ and b are constants. These constants can be written as $m = [u^*(x_1) - u^*(x_2)]/[u(x_1) - u(x_2)]$ and $b = [u^*(x_1)u(x_2) - u(x_1)u^*(x_2)]/[u(x_1) - u(x_2)]$ for any real numbers x_1 and x_2 with $u(x_1) > u(x_2)$.

(A)–(C) tell us that any utility function that yields the right fair prices for wagers *cardinally measures* the value of money. This means, first, that its expected values are *ordinally significant*, so that $U(G) > U(H)$ holds only if all rational professional bettors really do prefer holding G to holding H. Second, *ratios of differences* of these expected values have *cardinal significance*; they accurately express the amount by which various changes in a bettor's portfolio of wagers would contribute to her happiness. One way to put this is to say that the identity $U(G_1) - U(G_2) = \lambda[U(G_3) - U(G_4)]$ holds only when the change in happiness caused by gaining G_1 and losing G_2 is λ times the change caused by gaining G_3 and losing G_4. Last, ratios of u's expected values lack cardinal significance, so that $u(G) = 2u(G^*)$ cannot be taken to mean that G is twice as desirable as G^*. In light of all this, we may rewrite Theorem 1.2 as follows: *If a professional gambler's fair prices satisfy Premises 1–7, then there will be an interval scale, unique up to positive linear transformation, that quantifies the "value" of* money for her, and she will price wagers according to their expected "values" as measured on this scale.

We will discuss the concept of utility more fully in the next section, but we first must understand why it did not enter into Pascal's thinking. Pascal implicitly treated differences in the sizes of fortunes as differences in the amount of happiness they can buy. So, while the letter of his position commits him to all the premises used in our argument thus far, his conclusion requires one more to ensure that $u(x) = x$. The further premise simply says that Pascal's Thesis holds in the special case of an even odds bet.

Premise 8 (Risk Neutrality). *$(x + y)/2$ is the fair price for a wager that pays $x or $y with equal probability.*

In the presence of the other premises this turns out to be enough to ensure that fair prices go by expected payoffs generally.

Premise 8 can be given a plausible sounding rationale. Consider three wagers of the form

	p	$1 - p$
G_0	\$x	\$y
G_1	\$(x + y)/2	\$(x + y)/2
G_2	\$y	\$x

Cover up the G_2 row and observe that the difference between the payoffs of G_0 and G_1 in the first column is $(x - y)/2$, and in the second it is $-(y - x)/2$. Now cover the G_0 row and note that the difference between the payoffs of G_1 and G_2 in the first column is $(x - y)/2$, and in the second is $(y - x)/2$. Thus, the change in the agent's fortune in moving from G_0 and G_1 is the same as the change in her fortune in moving from G_1 and G_2 under both the p-event and the $(1 - p)$-event. Given this symmetry, it is reasonable to think that the difference in value between G_0 and G_1 should match the difference in value between G_1 and G_2. Accordingly, the fair prices should line up so that $g_1 =$ $(x + y)/2$ falls halfway between g_0 and g_2. Now, in the case where $p = 1/2$ we know that g_0 and g_2 will be identical (by Stochastic Equivalence), so $g_0 = g_2 = g_1 = $(x + y)/2$ as Premise 8 requires.

The fallacy here lies in the step from "the change in the agent's *fortune* in moving from G_0 and G_1 is the same as the change in her *fortune* in moving from G_1 and G_2" to "the difference in *value* between G_0 and G_1 should match the difference in *value* between G_1 and G_2." This inference goes through only if we assume that

31

Money Has Constant Marginal Utility. For any real numbers $x > y \geq 0$, the difference in desirability to a rational professional gambler between having a fortune of $\$y$ and having one of $\$(x + y)/2$ should be the same as the difference in desirability between having a fortune of $\$(x + y)/2$ and having one of $\$x$.

Among other things, this entails that a rational gambler should always desire a fortune of $\$2f$ twice as strongly as a fortune of $\$f$, and that her desire for an extra dollar should never wax or wane with changes in her fortune, so that the difference in desirability between $\$f$ and $\$(f + 1)$ is the same for every f. If this does not seem any more obvious to you than Premise 8, do not be alarmed – the two principles are equivalent in the context of the other premises, and neither is true.

1.5 THE CRAMER/BERNOULLI THESIS

Gabriel Cramer and Daniel Bernoulli deserve the credit for discovering that money does not have constant marginal utility (though Cramer's claim is often ignored).[15] It was Cramer who first realized that fair prices cannot plausibly be identified with expected payoffs. He came to this conclusion during the course of investigations into a gambling problem proposed by Daniel Bernoulli's brother, Nicholas, which came to be called the *St. Petersburg Paradox*. Suppose a fair coin is going to be tossed until a tail comes up, and let E_k be the event of the first tail coming up on the kth flip. Consider a wager of the form $G_f = \&_k(E_k \; \Box \rightarrow \$(f + 2^k))$, for $k = 1, 2, 3, \ldots$ and $f > 0$, where $\$f$ is your present net worth. If you hold G_f your net worth will increase by $\$2$ if the first tail comes up on the first toss, by $\$4$ if the first tail comes on the second toss, of $\$8$ if the first tail comes on the third toss, and so on. What do you think G_f's fair price ought to be? That is, at what value $g_f = (f + \delta_f)$ would you be equally happy having $\$g_f$ as your net worth or letting your fortunes ride on G_f? When asked this question most people set $\$g_f$ at roughly $\$20$ above their current fortune. That is, they profess to be indifferent between having an extra $\$20$ in their pocket or holding a bet that offers them an extra $\$2$ with probability 1/2, an extra $\$4$ with probability 1/4, an extra $\$8$ with probability 1/8, and so on. There are variations, of course, but nearly everyone chooses a low number, far lower than, say, $\$100,000$. This is not surprising since at $g_f = (f + 100,000)$ the chance of G_f paying more than $\$g_f$ is about 0.0000076, and the chance of it paying at least $\$90,000$ *less* than $\$g_f$ is 0.9998.

[15] For an excellent discussion of the history see Todhunter (1865/1949, pp. 213–38).

32

The trouble is that a person who identifies g_f with $\text{Exp}_\$(G_f)$ should think it wise to forgo *any* finite, risk-free increase in wealth to obtain G_f because G_f has an *infinite* expected payoff: $\Sigma_k \rho(E_k)(f + 2^k) = f + \Sigma_k (1/2^k)2^k = \infty$. She should, for example, prefer playing the St. Petersburg game to having a straight payment of a *trillion* dollars even though she is fully aware that the coin would need to come up heads *forty* times in a row for her to win that much. (To put this in perspective, the odds of tossing forty straight heads with a fair coin have about the same order of magnitude as the odds of a large asteroid hitting the earth in the next second.) Pascal's Thesis thus leads to patently absurd results when applied to the St. Petersburg game.[16]

The root of the problem, as Cramer recognized, is that the identification of fair prices with expected payoffs only makes sense if money has constant marginal utility. In the St. Petersburg game Pascal's Thesis requires a rational bettor to regard a potential increase of $\$2^{k+1}$ in her fortune as being twice as desirable as an increase of $\$2^k$. If she did not, there would be no reason to think that the former prospect should have the same effect on G_f's value when discounted by a probability half as large. This is just a special case of the general claim, inherent in the idea that money has constant marginal utility, that the difference in desirability between having a fortune of $\$y$ (read $\$f$) and having one of $\$(x + y)/2$ (read $\$(f + 2^k)$) should be the same as the difference in desirability between having a fortune of $\$(x + y)/2$ and having one of $\$x$ (read $\$(f + 2^{k+1})$). This was Pascal's mistake.

Money, contrary to the old adage, can buy happiness[17] (even if there are many things that contribute to happiness that it cannot buy), but the "rate of exchange" between the two is not constant. The "happiness value" of a given dollar depends on how many others one will have left once it is spent; money is always worth more to a pauper than to a prince. By identifying fair prices with expected payoffs Pascal was implicitly denying this. On his view the value of an extra dollar should be invariant among all professional gamblers *no matter how large or*

---

[16] A determined proponent of the Thesis might try to respond by suggesting that the (fair price = expected payoff) equation only applies to wagers based on *finite* event partitions or, at most, to infinite wagers with expected payoffs that converge to some finite value. This dodge will not work. The paradoxical character of the St. Petersburg game has nothing special to do with the fact that its expected payoff is infinite. The advice Pascal's Thesis offers is just as absurd in a "truncated" St. Petersburg game that has a finite upper limit of, say, one hundred thousand tosses (where it is stipulated that a bettor's fortune stays at $\$f$ if all one hundred thousand come up heads).

[17] I am not using *happiness* here as the name of a pleasurable or satisfied *feeling*, but in the broader sense in which happiness encompasses all aspects of (subjective) well-being, the way the term is used in "1987 was the happiest year of my life."

small their fortunes happen to be. Cramer recognized this as a mistake; the "amount of happiness" that $\$2^{k+1}$ can buy is usually less than twice the amount $\$2^k$ can buy, far less when k gets large. In effect, then, Cramer rejected the idea that money has constant marginal utility, and with it Premise 8 of our argument for Pascal's Thesis.

Once one does this a way of resolving the St. Petersburg Paradox suggested itself almost immediately. If the "value" of having $\$2^k$ rises slowly enough as k grows, then there can be a finite amount of money that would buy the same "amount" of happiness as G_f can be expected to buy. This would make g_f finite even though G_f's expected payoff is infinite. Of course, for this solution to work there needs to be a way of making sense of talk about the "value" of money or the "amount" of happiness it buys. Thus it was that Cramer first hit upon the idea of a real function $u(x)$ that would measure the overall desirability of having a fortune of $\$x$. Let's call Cramer's function a "utility" even though this term was not used until long after his death. (Also, for historical accuracy, it should be noted that Cramer saw u as measuring the value of *changes* in a bettor's net worth rather than the value of her total fortune.)

Once one starts thinking in terms of utility rather than money it is a short step to the view that each wager should be valued by its *expected utility* $U(G) = \int_S u(g_s)d\rho$ (or $U(G) = \Sigma_S u(g_s)\rho(S)$ when the set of states is countable) rather than its expected payoff. Cramer thus took $U(G)$ to be the right measure of the "amount of happiness" that a gambler can reasonably expect to gain from holding G. This led him to replace Pascal's Thesis by

The Cramer/Bernoulli Thesis. *If the utility $u(x)$ accurately measures the amount of happiness or satisfaction that a fortune of $\$x$ can buy, if $U(G)$ is the expected utility of the wager G (computed relative to ρ), and if $\$g$ is a sum of money such that $u(g) = U(G)$, then any professional gambler should find the prospect of holding G and the prospect of having a fortune of $\$g$ equally desirable.*

If this is right, the St. Petersburg Paradox can be solved simply by showing that the correct utility function for money makes $\Sigma_k 1/2^k u(2^k)$ finite even when $\Sigma_k (1/2^k)2^k$ is infinite. Cramer knew that $u(2^{k+1}) < 2u(2^k)$ was a sufficient condition for $U(G_f) < \infty$,[18] and he maintained

---

[18] This might not be obvious. Since the terms of the series $\Sigma_k 1/2^k u(2^k)$ can be chosen to be positive one can appeal to a version of the *ratio test*: If $a_k > 0$ for $k = 1, 2, 3, \ldots$, then $\Sigma_k a_k$ converges if $a_{k+1}/a_k < 1$ for all k. Applying this with $a_k = u(2^k)/2^k$ yields the desired result.

that any reasonable utility should obey the former inequality. He did not go on to defend any specific u as *the* correct utility function, though he did point out that the expected utility of the St. Petersburg game would be finite when $u(\$x) = x^{1/2}$ or when $u(2^k)$ is constant for all $k > 24$.

By requiring that $u(2^{k+1}) < 2u(2^k)$ Cramer was, in effect, appealing to the fact that

Money Has Declining Marginal Utility.[19] *For any real numbers $x > y$, the difference in desirability, from a professional gambler's point of view, between having a fortune of $\$x$ and having one of $\$(x + y)/2$ should be greater than the difference in desirability between having a fortune of $\$(x + y)/2$ and having one of $\$y$.*

to provide a rationale for replacing Premise 8 of the previous section by

Premise 8* (Risk Aversion). *The fair price for a wager that pays $\$x$ and $\$y$ with equal probability should be strictly less than $\$(x + y)/2$.*

Given the other premises, Premise 8* implies that the gambler should be *risk averse* at all wealth levels. In the jargon of modern economics, a person is *risk averse*, *risk neutral*, or *risk seeking* at wealth level $\$f$ just in case there is positive number ε such that for any $0 < \delta < \varepsilon$ she would find having a sure fortune of $\$f$ strictly more, exactly as, or strictly less desirable than having the wager

	1/2	1/2
$G(f, \delta)$	$\$(f + \delta)$	$\$(f - \delta)$

A risk averter will set a price $\$g(f, \delta)$ for $G(f, \delta)$ that is less than its expected payoff $\$f$, a risk neutralist will set $\$g(f, \delta) = \$f$, and a risk seeker will set $\$g(f, \delta) > \$f$. Pascal's Thesis gives absurd results because it assumes that all rational bettors are risk neutral everywhere. Cramer's great insight was to see that the hypothesis of universal risk neutrality should be replaced by the hypothesis of universal risk aversion.

---

[19] There are few economic hypotheses that have more explanatory power than the declining marginal utility of money. It accounts for the willingness of poor people to work harder for a dollar than most rich people would ever dream of; the tendency of the wealthy to spend a higher fraction of their income on "luxuries" and their willingness to invest more heavily in risky ventures (including education). It also explains the existence of institutions such as the insurance industry, trading in common stocks and futures, sharecropping, sweatshops, and long-term labor contracts.

Daniel Bernoulli gets part of the credit for the Cramer/Bernoulli Thesis because he was first to suggest a plausible candidate for u, first to appreciate that utilities attach to wealth levels rather than changes in wealth, and first into print. Bernoulli, who ranks among the foremost mathematicians of the eighteenth century (even though he was only mediocre for a Bernoulli), learned of the St. Petersburg problem by reading a letter from Cramer to his *cousin* Nicholas Bernoulli[20] outlining the expected utility hypothesis. Daniel immediately recognized the merit in Cramer's proposal and went beyond it by arguing that utility must be a *logarithmic* function of money. He arrived at this conclusion by supposing that the utility gained by increasing a fortune of $\$f$ by an increment of $\$\Delta f$ should always be inversely proportional to the fortune's size and directly proportional to that of the increment, so that $u(f + \Delta f) - u(f) = \alpha \Delta f / f$ for some positive α. The only functions with this property have the logarithmic form $u(x) = \alpha\log(x) + \beta$, where $\alpha > 0$ and β are real constants. These provide a system of interval scales for measuring the value of money with different choices of α and β indicating different conventions for setting the unit and zero. As was noted in the previous section all such scales determine the same fair prices, so we lose nothing by simply setting $u(x) = \log(x)$.

This function is risk averse everywhere,[21] and it produces finite fair prices for the St. Petersburg game that seem fairly reasonable. Here is a sampling:

Value of f	Fair price for G_f	Amount by which $\$G_f$ exceeds $\$f$
$1,000	$1,012	$12.00
$2,000	$2,013	$13.00
$10,000	$10,015.25	$15.25
$20,000	$20,016.25	$16.25
$100,000	$100,018.50	$18.50
$200,000	$200,019.50	$19.50
$1,000,000	$1,000,021.50	$21.50
$2,000,000	$2,000,022.50	$22.50

Notice that the fair price of G_f increases with f. This corresponds to the often observed fact that the *degree* of risk aversion tends to decline with increases in an agent's fortune. (The rich, in other words,

---

[20] This Nicholas is not to be confused with Daniel's *brother* Nicholas, who posed the St. Petersburg problem to Cramer.

[21] Proof: $[\frac{1}{2}\log(f + \delta) + \frac{1}{2}\log(f - \delta)] = \frac{1}{2}\log(f^2 - \delta^2) < \frac{1}{2}\log(f^2) = \log(f)$.

tend to be more willing than the poor to risk wealth in speculative ventures.)

While modern economists follow Cramer and Bernoulli in thinking both that money has declining marginal utility and that wagers should be valued by their expected utilities, they reject Bernoulli's claim that $u(x) = \log(x)$. The fact is that Bernoulli's definition does not really solve the St. Petersburg Paradox, but merely sweeps it under the rug. To see why, consider the wager $H_f = \&_j (E_j \square \rightarrow (f + \$\exp(2^j)))$, where $\exp(x)$ is the base of the natural logarithm raised to the xth power. On Bernoulli's proposal H_f has infinite expected utility. The problem with this is that anyone who assigns H_f infinite expected utility will assign infinite utility to any arrangement $H_\rho = \rho H_f \oplus (1 - \rho)\$x$ in which she stands to gain H_f with some probability $\rho > 0$ (no matter how small) and a finite "consolation prize" of $\$x$ with probability $1 - \rho$. This causes a number of formal problems. For instance, Stochastic Dominance is violated because, even though it seems that a bettor holding H_ρ ought to do better as ρ gets larger, this is not borne out by the expected utilities, which are all infinite. Second, a wager of infinite utility will be strictly preferred to *any* of its payoffs since the latter are all finite. This is absurd given that we are confining our attention to bettors who value wagers only as means to the end of increasing their fortune.

All this aside, the real problem is that the infinite utility of H_f forces a decision maker completely to ignore facts about what might happen if she does not obtain H_f, even though these facts clearly should be relevant to her deliberations. No matter how close ρ is to zero, an expected utility maximizer will be unable to see any difference between a case in which the consolation prize is abject poverty ($x = 0$) and one in which it is fabulous wealth ($x = \$10,000,000$) since both *total* situations will seem infinitely desirable to her. Thus, an agent who has a chance at some infinitely desirable outcome does not need to consult her beliefs or other desires in coming to a decision about what to do or want. Her task is clear and simple: She should try to obtain the infinitely desirable thing by any means that affords her some chance of getting it, and *any one of these means is just as good as any other*. She must, in short, become blind to "merely finite" things in her pursuit of the infinite good. This is clearly irrational when a "merely finite" thing is all one is likely to get.

To see the point most vividly, consider Pascal's famous "wager" argument, which tries to use expected utility theory, in a qualitative way, to convince people to go to church, take the sacraments, and lead upright lives. The claim is that doing these things is likely to lead to belief in God, which increases one's chances of securing the infinite rewards of paradise. There are many well-known criticisms of Pascal's reasoning, but the most telling is one due to Anthony Duff and Alan Hajek.[22] While neither Duff nor Hajek puts things quite this way, the basic point is that Pascal's conclusion does not follow from his premises because even if going to church, taking the sacraments, and living an upright life did give one a chance at infinite happiness, this would not be a reason to prefer such a life to one that involves torturing and killing those who go to church, take sacraments, and live uprightly because, as the case of St. Paul amply demonstrates, there is a nonzero probability that this too will secure one a place in the New Jerusalem. It would do no good for Pascal to respond that the probability of eternal bliss is higher for do-gooders than for those who persecute them. After all, by his own lights, the magnitudes of these probabilities become irrelevant when they are multiplied by an infinite utility. This, indeed, is the very point on which his argument hinges.

There are three ways to prevent wagers from having infinite fair prices so as to avoid any version of the St. Petersburg Paradox. The simplest is to claim that the expected utility analysis of fair prices makes sense only when it is applied to wagers based on *finite* event partitions. This solves the problem, but only at the expense of imposing a limitation on decision theory that seems far too extreme. It is perfectly possible, after all, for us to make sense of wagers that have infinitely many possible outcomes. As a second option, one might allow unbounded utility functions, but restrict the allowable gambles in such a way that events of high utility are always assigned such low probabilities that infinite expected utilities never arise.[23] The St. Petersburg Paradox would not be a problem on this view because there would be no chance that anyone would ever be faced with it. A third approach, the one I prefer, is to require that a rational agent's utility function for money be *bounded* in the sense that there is a $b > 0$ such that $-b < u(\$x) < b$ for all $x \in [0, \infty]$. Under these conditions every wager the agent considers will have a finite expected utility, and the sorts of problems we have been discussing will not arise. Since $\alpha \log(x) + \beta$ fails all three of these tests it cannot be the correct utility function for money (though it may be a good approximation in regions far from infinity). This is one reason for rejecting Bernoulli's definition of utility. There is another.

---

[22] I first heard this criticism of Pascal's wager (albeit in a slightly different form) from Alan Hajek. His is the most sophisticated treatment of the wager I have yet seen. The basic point had been made earlier by Duff in his (1986).

[23] This can be done consistently. For details see Kreps (1988, pp. 63–68).

A second reason for rejecting Bernoulli's definition is that it is wrong to think that any *one* definition of utility should hold good for everyone. There is no *single* right way for a professional gambler to value money. The desire for money is (to some extent) a matter of taste, and some people have more of a taste for it than others. Since rational choice theory cannot be in the business of dictating tastes, it should start by taking a gambler's desire for money as a datum, and go on to dictate *her* fair prices in terms of it. From the perspective of the theory there is no such thing as *the* correct utility function for money, *the* right fair price of a wager, or *the* right risk premium to pay at any level of wealth. A professional gambler's desire for money is ultimately an instance of *de gustibus non disputandum*.

Even though Bernoulli was wrong to insist that each rational gambler must have exactly the same utility function, he was right to think that there should always be some utility function or other that accurately measures the "amount of happiness" that fortunes of various sizes will buy *for her*. While this would make it incorrect to say that a gamble $G = \&_S (S \,\square\!\!\rightarrow \$g_S)$ has *the* fair price given by $\log(\$g) = \Sigma_S \rho(S)\log(\$g_S)$, it would remain true that for each rational bettor there must be some number that is G's fair price *for her*. So, in backing away from Bernoulli's definition, one might still want to accept the Cramer/Bernoulli Thesis subject to the proviso that the appropriate utility function, and the fair prices associated with it, will vary from agent to agent.

This leaves us with the problem of saying which utilities go with which agents. Let us think of each gambler as being characterized by a set of statements that describe the strengths of her desires for wealth, and wagers involving wealth. We will call these statements *constraints*[24] because they constrain the form of the utility function that is used to measure the strength of the agent's desires. The most common constraints, and the ones that will be most important in the present context, specify the strengths of the gambler's desires in *comparative* terms by saying things like the following:

- She strictly prefers holding G to holding G^*.
- Her fair price for G is $\$g$; that is, she is indifferent between holding G and having a guaranteed fortune of $\$g$.
- She is everywhere risk averse.

---

[24] I am taking this use of the term *constraint* from van Fraassen (1984).

Other constraints have a *cardinal* character. They might say things like the following:

- The difference between the strength of the gambler's desire for G and the strength of her desire for G^* is twice the difference between the strength of her desire for H and the strength of her desire for H^*.

A utility u will be said to *represent* the bettor's desires if anyone who formed desires about wagers by maximizing expected u values would automatically satisfy all the constraints that the bettor's desires satisfy. Thus to satisfy the constraints listed above we would need a utility u for which

- $U(G) > U(G^*)$
- $u(\$g) = U(G)$
- $u(\$f) > U(G(f, \delta))$ for all f and small δ
- $U(G) - U(G^*) = 2[U(H) - U(H^*)]$

Once one rejects the notion of a one-size-fits-all utility function the next natural step is to require that every rational agent's fair prices be described by a utility that represents her desires.

Decision theorists have typically been concerned with ordinally specified constraints on desire, what are often called *preferences*. We think of these as an agent's "all-things-considered" judgments about the desirability of wagers in **G**. Since **G** contains all the *constant* wagers that pay out the same fixed sum $\$x$ under every contingency, the desirability of riskless as well as risky prospects can be modeled as preferences for wagers in **G**. A bettor might be in any of the following six attitudinal states vis-à-vis the comparative desirability of any pair of wagers G and G^*:

- G is *strictly preferred* to G^* (written $G > G^*$); from the agent's perspective it would be more desirable to hold G than to hold G^*.
- G is *weakly preferred* to G^* ($G \geq G^*$); from the agent's perspective it would be at least as desirable to hold G as to hold G^*.
- The agent is *indifferent* between G and G^* ($G \approx G^*$); the agent finds the prospect of holding G or holding G^* equally desirable.
- G^* is *strictly preferred* to G ($G^* > G$).
- G^* is *weakly preferred* to G ($G^* \geq G$).
- *No preference*: The strengths of the agent's desires regarding G and G^* are not sufficiently determinate to warrant any of the previous characterizations.

40

The totality of these attitudes is the agent's *preference ranking* over wagers.[25] Formally, this is a pair of binary relations ($>$, $\geq$) defined on **G** in which $>$ encodes the agent's strict preferences and $\geq$ encodes her weak ones. Indifference can be introduced via the definition: $G \approx G^*$ if and only if $G^* \geq G^*$ and $G^* \geq G$. The agent's fair prices show up within ($>$, $\geq$) as indifferences of the form $\$g \approx G$ where $\$g$ is thought of as a constant wager that pays $\$g$ in each state of the world.

If we think of each agent's desires as being fully described by her preference ranking, then one natural way to apply the expected utility hypothesis is by making it a requirement of instrumental rationality that there should always be an agent-specific *subjective utility function* u defined on **G** that *strongly ordinally represents* her preferences in the sense that, for any two wagers G and G^* in **G**, the agent strictly (weakly) prefers G to G^* if and only if G's expected utility is greater than (not less than) G^*'s utility where both these utilities are computed using u. Formally, the requirement would be as follows:

Subjective Expected Utility Hypothesis (strong version). *In order for a "professional" gambler to be rational in the instrumental sense there must be at least one utility function for money, u, whose associated expectation operator, U, strongly ordinally represents the gambler's preference ranking in the sense that both*

$$G > G^* \text{ if and only if } U(G) > U(G^*) \text{ and}$$
$$G \geq G^* \text{ if and only if } U(G) \geq U(G^*)$$

hold for any G, $G^* \in$ **G**.

On the reasonable assumption that an instrumentally rational bettor will always perform her most preferred option, this principle entails that such a bettor will invariably seek to obtain a wager that maximizes her subjective expected utility. Moreover, since **G** contains all "constant" wagers, and since each $G \in$ **G** can be associated with a unique sum $\$g$ at which the bettor is indifferent between $\$g$ and G, the identity $u(\$x) = U(G)$ will hold just in case $x = g$. Thus, the fair price

---

[25] The terminology I am using is somewhat nonstandard. Usually the phrase "preference ranking" is taken to denote a *complete ordering* of **G** in which the "no preference" alternative never occurs, so that the *trichotomy property* – $G > G^*$ or $G \approx G^*$ or $G < G^*$ – holds for every pair G, $G^* \in$ **G**. However, since we are going to need a convenient term to denote both incomplete order relations, or *partial* orderings, as well as complete orderings, I am going use "ranking" to refer to both. This should cause no confusion so long as readers keep in mind that *rankings need not be complete*.

for any wager will conform to its expected utility, just as Pascal, Cramer, and Bernoulli suggested, but it is now the bettor's *subjective* expected utility that matters.

The classic defense of the Subjective Expected Utility Hypothesis was presented by John von Neumann and Otto Morgenstern in their immensely influential *Theory of Games and Economic Behavior*.[26] The argument they give is a generalization of Theorem 1.2 presented earlier. Von Neumann and Morgenstern set down a set of axioms on rational preference rankings whose satisfaction was sufficient to ensure that the ranking could be represented by an expected utility function. Here is a system of axioms that is equivalent to the set they presented:[27]

VN$_1$ ($>$, $\geq$) *partially orders* **G**.

> *Reflexivity of Weak Preference.* $G \geq G$.
> *Consistency.* $G > G^*$ only if $G \geq G^*$
> $\qquad\qquad G \geq G^*$ only if not $G^* > G$
> *Transitivity.* If $G \geq G^*$ and $G^* \geq G^{**}$, then $G \geq G^{**}$
> $\qquad\qquad$ If $G > G^*$ and $G^* \geq G^{**}$, then $G > G^{**}$
> $\qquad\qquad$ If $G \geq G^*$ and $G^* > G^{**}$, then $G > G^{**}$

VN$_2$ *Averaging.* If G's lowest payoff is at least as high as G^*'s highest payoff, then $G \geq G^*$. If, in addition, there is a nonzero probability that G and G^* have different payoffs, then $G > G^*$.

VN$_2$ *Independence.* $G > G^*$ if and only if $G \oplus G^{**} > G^* \oplus G^{**}$
$\qquad\qquad\qquad G \geq G^*$ if and only if $G \oplus G^{**} \geq G^* \oplus G^{**}$

VN$_3$ *Stochastic Dominance.* Let G pay x_1 with probability ρ and $x_2 < x_1$ with probability $1 - \rho$ and let G^* pay x_1 with probability ρ^* and x_2 with probability $1 - \rho^*$. Then, $G \geq G^*$ iff $\rho \geq \rho^*$.

VN$_4$ *Archimedes' Axiom.* If $G > G^* > G^{**}$, then there is a real number λ strictly between 0 and 1 such that $\lambda G \oplus (1 - \lambda)G^{**} > G^* > (1 - \lambda)G \oplus \lambda G^{**}$.

VN$_5$ *Linearity:* For any real number $\lambda > 0$, $G \geq G^*$ if and only if $\lambda G \geq \lambda G^*$, and $G > G^*$ if and only if $\lambda G > \lambda G^*$.

VN$_6$ *Completeness:* Either $G > G^*$ or $G^* \geq G$.

---

[26] The idea of using utilities to represent preferences had appeared earlier in the writings of Frank Ramsey, but his work was not well known, and not very influential, until after the publication of *Theory of Games*.

[27] I have altered the von Neumann/Morgenstern axioms merely to highlight the analogies with Premises 1–7.

Von Neumann and Morgenstern showed how these axioms can be used to construct an expected utility representation for any preference ranking that satisfies them. Here is their result:

Theorem 1.3 (von Neumann/Morgenstern). *If a preference ranking* $(>, \geq)$ *satisfies* $VN_1 - VN_6$, *then there exists a utility function for money* u *whose associated expectation operator* $U(G) = \Sigma_S \rho(S) u(g_S)$ *strongly ordinally represents* $(>, \geq)$ *in the sense that*

$$G > G^* \text{ if and only if } U(G) > U(G^*)$$
$$G \geq G^* \text{ if and only if } U(G) \geq U(G^*)$$

for any wagers $G, G^* \in$ **G**. *Moreover, u is unique up the arbitrary choice of a unit and a zero point relative to which utility is measured.*

Theorem 1.3 tells us that a preference ranking defined over **G** can only satisfy the von Neumann/Morgenstern axioms if it can be represented by a *cardinal* utility function whose expected values measure the "amount of overall happiness" that the wagers in **G** can buy. Thus, anyone who holds that all of the von Neumann/Morgenstern axioms are strict requirements of instrumental rationality is thereby committed to the view that the strengths of a rational bettor's desires can be measured on an interval scale with the same degree of precision that the temperatures of physical objects can be measured on the Celsius scale.

This is a strong requirement. Indeed, most proponents of expected utility maximization now think it is too strong. It commits what Mark Kaplan has called "the sin of false precision"[28] by supposing that the desires of any rational gambler will be sufficiently precise to be measured by real numbers. Since her preference ranking must be represented by a *single* utility function (once a unit and a zero are in place) there is no room for any kind of vagueness or indeterminacy in the strengths of her desires. Just as every physical object is taken to have a definite temperature, so every desire must have a definite strength. This is clearly unrealistic. Many desires lack determinate strengths, and this is no indication of the irrationality of those who hold them. Nonmonetary cases provide the most compelling examples: If hearing no music is 0 and hearing Mozart's *Jupiter Symphony* played by a first-class orchestra is 1, how much would it be worth, in terms of your overall happiness, to hear John Coltrane's *A Love Supreme* played by

[28] Kaplan (1994, pp. 23–31).

43

an excellent jazz quartet? Your tastes will probably dictate whether the answer should be "more than 1," "less than 0," "somewhere between 0 and 1," "closer to 1 than 0," or even, "very close to 1." I doubt, however, that you can come up with a number that gauges the strength of your desire to the third decimal place. This is not because you don't know what the number is; it is because there is no such number to be known. Even in the case of wagers with monetary outcomes it seems unlikely that our desires are anywhere near as definite as the von Neumann/Morgenstern axioms demand. Exactly how much is a fortune of $100,000 worth if $0 has utility 0 and $200,000 has utility 1? Or, at precisely what value of ρ would you be indifferent between having a sure $10,000 and a wager that pays $20,000 with probability ρ and $0 with probability $1 - \rho$? It is no sin against rationality not to know how to answer these questions, or for them not to have any answers at all.[29] The general point here is that once we give up on finding a one-size-fits-all utility we open up space for the idea that people can differ not only in what they want and how strongly they want it, but in the extent to which these wants are determinate.

Indeterminacy in a bettor's desires shows up as *incompleteness* in her preferences; the trichotomy property – $G > G^*$, $G \approx G^*$, or $G < G^*$ – will fail for certain (G, G^*) pairs. Whenever it does, there will be wagers that lack fair prices; that is, for some $G \in \mathbf{G}$ the agent's preference ranking will not contain *any* indifference of the form $\$x \approx G$, for x a real number. In such cases, one can still employ talk of G's fair price to convey information about the agent's preferences, using things like, "G's fair price lies between $8,000 and $9,000," to mean that she strictly prefers (disprefers) G to every sum less (greater) than $8,000 ($9,000). In speaking this way, however, we must be careful not to slip into what Isaac Levi calls "black box" thinking, which treats any inexact statement about a wager's fair price for a gambler as nothing more than an expression of our ignorance of its true price for her.[30] In cases where the strengths of her desires are genuinely indeterminate there will be wagers whose fair price we

---

[29] A bad reply here would be to suggest that your utility for $100,000 just *is* that number you would *choose* if you had to pick a value for ρ at which you would get either the money or the wager with equal probability. Aside from taking us back to the behaviorist view of desire rejected above, this assumes, falsely, that the choice you make will always reveal your degree of desire. It probably would if you had one, but it is perfectly consistent with having no definite degree of desire that you choose a precise value for ρ when forced to do so. This is especially evident when the choice is not underwritten by any *stable* behavioral disposition.

[30] Levi (1985).

cannot know, not because our access to her mental states is somehow limited, but because there is nothing to be known. More generally, we must take care not to regard the incompleteness in a rational gambler's preference ranking as indicating gaps in our (or her) knowledge about the true interval scale that measures the strengths of her desires. A gambler can be rational even if her beliefs are too vague to be measured on any single interval scale. We must, therefore, reject the idea that the von Neumann/Morgenstern axioms are all necessary conditions for instrumental rationality since such a view would falsely entail that there can be no vagueness in preferences.

The problematic axiom is the completeness principle VN_6. Many proponents of expected utility theory[31] now interpret this axiom not as a law of rationality per se, but as a *requirement of coherent extendibility*. We will discuss the implications of this at some length in Chapter 3. The basic idea is that VN_6 should be replaced by

VN_6^+ *Coherent Extendibility*: While ($>$, $\geq$) need not completely order **G**, there must be at least one (complete) preference ranking ($>^+$, $\geq^+$) that both satisfies the axioms $VN_1 - VN_6$ and *extends* ($>$, $\geq$) in the sense that $G > G^*$ implies $G >^+ G^*$ and $G \geq G^*$ implies $G \geq^+ G^*$.

In other words, it should always be possible in principle to "flesh out" a rational agent's preferences (usually in more than one way) to obtain a complete ranking that does not violate any of the von Neumann/Morgenstern axioms.

When VN_6 is replaced by VN_6^+ the expected utility theorist's basic requirement of rationality becomes

Subjective Expected Utility Hypothesis (weak version). *In order for a "professional" gambler to be rational in the instrumental sense there must be at least one utility function for money, u, whose associated expectation operator, U, weakly ordinally represents the gambler's preference ranking in the sense that*

$$G > G^* \text{ only if } U(G) > U(G^*)$$
$$G \geq G^* \text{ only if } U(G) \geq U(G^*)$$

*hold for any G, $G^* \in$ **G**.*

[31] See, for example, Kaplan (1989) and (1996), and Jeffrey (1983).

This, finally, is the principle that expresses the basic truth underlying Pascal's insight: It sets up *representation of desires by some utility function or other* as the basic criterion of instrumental rationality. In does not, however, ask that there be a single utility that correctly measures the strength of an agent's desires (up to the choice of a unit and zero point), nor does it force the agent to have a definite fair price for every wager she considers. Nonetheless, it preserves the essential core of the Pascalian position by making expected utility maximization central to rational betting, and by retaining the idea that there should always be at least one (but usually many) utility functions for which it is true $u(\$g) = U(G)$ in any case in which an agent's desires are sufficiently definite to determine that $\$g$ is her fair price for G.

1.7 LIFE OUTSIDE THE CASINO

In my view, the weakened version of the Subjective Expected Utility Hypothesis expresses a correct norm of instrumental rationality for "professional" bettors who are evaluating risky wagers. This is, of course, a rather limited claim; it is only applicable to decisions made in the casino when money is at stake. The authors of the *Port Royal Logic* would not be pleased with us if we stopped here. They saw the new concepts of probability and expectation not merely as gambling aids, but as the foundation for a general methodology of right decision making. The final section of the *Logic*, in which the material on probability and expectation is found, has a distinctly apostolic ring to it. Its author, Antoine Arnauld (who either consulted Pascal directly or worked from notes left after his death in 1662), clearly felt that Pascal was on to something much more important than a recipe for successful gambling. He saw himself as presenting an approach to decision making that could be applied everywhere to improve people's lives. This is nowhere explicitly stated, but the examples tell the tale. An appreciation of the laws of probability and expectation is supposed to liberate the ignorant from irrational fears of being struck by lightning or other similar events that, though awful in their consequences, have such low probabilities that it is not worthwhile to worry about them. It will also help foolish princesses overcome their inability to walk into rooms for fear that the ceiling will fall on them. It can even be used to make the biggest decision of all, whether to take actions that would lead to a belief in God, which Pascal famously viewed as a kind of a wager.

Thus, the core tenet of utility theory was already in place more than four hundred years ago in the form of Pascal's Thesis that *rational decision making is a matter of using probabilities to calculate expected values for risky or uncertain prospects and of letting these expected values guide one's choices. Rational choice, in short, is governed by the laws of mathematical expectation.* There have been significant changes in decision theory since Pascal's time, but this basic insight remains at the core of all subsequent work.

The goal of future chapters is to see how this insight can be generalized and expanded to include more realistic decision making situations, such as those in which the "prizes" to be won are not merely monetary, and those in which the decision maker does not know the objective probabilities of all the events that are relevant to the outcomes of her choices. We need to start by getting clear about what decisions are.

2

Decision Problems

This chapter sets the stage for future developments by providing a formal characterization of the sorts of *decision problems* that agents face when choosing among alternative courses of action at a given time. The classic treatment of this subject, and the one still endorsed by most economists, can be found in Leonard Savage's masterpiece *The Foundations of Statistics.*[1] A somewhat different characterization of decision making situations, which tends to find favor with philosophers, has been offered by Richard Jeffrey in *The Logic of Decision.*[2] The model to be presented here will be a kind of hybrid of the Savage and Jeffrey approaches. It agrees with Savage in retaining a clear distinction among actions, states, and outcomes but follows Jeffrey in treating all three kinds of entities as species of *propositions*. The chief advantage of this mixed approach is that it allows us to be much more specific than Savage was about what the elements of decision problems actually are and how they relate to one another. It also gives us a general formal framework within which both evidential and causal decision theories can be conveniently expressed.

2.1 SAVAGE'S MODEL OF DECISION PROBLEMS

Savage's model envisions a *decision maker*, or *agent*, who chooses among risky or uncertain prospects, called *actions*, whose consequences or *outcomes*[3] depend on the *state of the world*. One describes the agent's *decision problem* by specifying a set **A** of possible acts among which she must decide, a class **O** of outcomes that provides an inventory of all the desirable or undesirable things that could befall the agent as a result of her choice, and a list **S** of states of the world

---

[1] Savage (1954/1972).
[2] Jeffrey (1965/1983).
[3] While Savage uses the term *consequences*, I prefer *outcomes*. The former has a distinctly causal flavor to it since *consequence* often just means *effect*. The less causally tainted term *outcome* is better in this context because, as we shall see, Savage's theory is consistent with either a causal or a noncausal interpretation. One should think of an outcome in this chapter as something that happens *in conjunction* with an action, not necessarily as something that happens as an *effect* of the action.

that describe possible external conditions that determine what outcome each act in **A** will produce. In cases were only finitely many acts, states, and outcomes need to be considered, such a problem can be represented by a matrix of the form:

STATES

		S_1	S_2	$S_3\dots$	S_n
	A_1	$O[A_1, S_1]$	$O[A_1, S_2]$	$O[A_1, S_3]\dots$	$O[A_1, S_n]$
A	A_2	$O[A_2, S_1]$	$O[A_2, S_2]$	$O[A_2, S_3]\dots$	$O[A_2, S_n]$
C	A_3	$O[A_3, S_1]$	$O[A_3, S_2]$	$O[A_3, S_3]\dots$	$O[A_3, S_n]$
T	.	.	.	.	.
S	.	.	.	.	.
	.	.	.	.	.
	A_m	$O[A_m, S_1]$	$O[A_m, S_2]$	$O[A_m, S_3]\dots$	$O[A_m, S_n]$

OUTCOMES

where $O[A, S]$ denotes the outcome that act A would produce if the state S were to obtain. For a concrete example of a decision problem, consider a woman who is deciding whether to walk or drive to work on a cloudy day. If this woman sees walking and driving as her only options, and if she thinks that the state of the weather is the only thing that might affect the outcome of her action, then she might well face the problem

	It rains.	It does not rain.
I walk to work.	I walk to work in the rain.	I walk to work, but not in the rain.
I drive to work.	I drive to work in the rain.	I drive to work, but not in the rain.

The point of factoring decision problems into these three components is to effect a kind of "division of conceptual labor" in which the states in **S** serve as the locus of all the agent's uncertainty; the outcomes in **O** provide the objects of her basic, noninstrumental desires; and the acts in **A** are the objects of her instrumental desires. While the decision maker is taken to know both her repertoire of possible actions and the outcomes that will accompany these acts in each possible state of the world, she typically will not know which state actually "obtains." This ignorance forces her to make estimates of the likelihoods of various states in **S** and to factor these in with her basic desires for outcomes in **O** in order to arrive at a view about which act in **A** provides the best means to the end of achieving a pleasing

49

outcome. Thus a woman trying to decide whether to walk or drive to work on a cloudy day will consult her beliefs about the likelihood of rain and her views about the relative merits of wet walks, wet drives, dry walks, and dry drives and will form some judgment about which of the two acts is more likely to lead to a desirable result.

Savage was not as clear as he might have been about what kinds of entities the elements of **A**, **O**, and **S** are supposed to be. Aside from presenting a few examples to give his readers a sense of what he had in mind, he was content to leave the notions of a state and a consequence as theoretical primitives. In his most explicit remarks on the topic, he wrote that a state "is a description of the world leaving no relevant aspect undescribed,"[4] whereas "a consequence is anything that might happen to the person."[5] He did say a bit more about actions, identifying them with abstract *functions* from states to outcomes. That is, he treated each $A \in \mathbf{A}$ as a map that assigns a unique outcome $A(S) = O[A, S]$ to each state S. This makes intuitive sense since acting often does set up a kind of functional relationship between contingencies and outcomes; the woman who chooses to walk to work on a cloudy day really does make her happiness a function of the weather. Nonetheless, knowing that an act is a function from **S** to **O** does not tell us much about it until we know what states and outcomes are. We will need a better understanding of these concepts to make headway against the sorts of problems we will be considering.

As Richard Jeffrey first observed, we can obtain a more fruitful and unified model of decisions by construing all three of their basic components as *propositions*.[6] The idea is to focus not on the events and conditions in the world that can influence a decision maker's happiness, but on the propositions that *describe* these events and conditions. On such a reading, states are propositions that describe external circumstances over which the agent has no direct control, and about which she typically has less than perfect knowledge. Here are some of the things that a state proposition might specify:

- whether a particular egg is rotten (Savage's example)
- whether a certain stock will go up or down this week
- whether the agent has a pack-a-day cigarette habit
- whether the agent will refrain from smoking tomorrow if she resolves to do so today

[4] Savage (1954/1972, pp. 9 and 13).
[5] Savage (1954/1972, p. 13).
[6] Jeffrey (1965/1983, pp. 59–85).

A state "obtains," as Savage has it, just when it's true. Outcomes, on Jeffrey's proposal, are propositions that describe circumstances that might make a difference to the agent's overall happiness. Her preferences are to be thought of as directed toward the truth-values of these outcome propositions. Instead of saying that a woman does not want to walk to work in the rain, for instance, we might say that she wants the proposition expressed by "I walk to work in the rain" to be false.[7] Jeffrey identifies acts with propositions whose truth-value the decision maker can directly control, those "which [are] within the agent's power to make true if he pleases."[8] Thus, choosing to walk to work is portrayed as choosing to make the proposition expressed by "I walk to work" true. While this is the right way to think of what one does in performing a deliberate action, we shall shortly see that it is not the best way to characterize the objects that Savage calls acts.

The principal advantage of Jeffrey's approach is that it allows us to apply standard logical operations to elements of decision problems, thus greatly extending the model's expressive power. Since propositions can be combined under Boolean operations, we can speak of the disjunctive action $A \vee A^*$ that the agent "performs" either by doing A or by doing A^*. Likewise, we can conjoin an act A and a state S to get $A \& S$, the proposition that A is performed and S obtains. We can even express propositions like $(A \& S) \vee (\neg A \& \neg S)$, which is the sort of thing an agent might believe if he took there to be a high degree of correlation between the performance of A and the occurrence of S. (This kind of case will figure in the sequel.) Equally important is the fact that we can form subjunctive conditionals involving acts, states, and outcomes to obtain statements like $S \,\Box\!\!\rightarrow\, O$, which says that the agent would receive O if the S were to obtain. Or, we can write $(A \,\Box\!\!\rightarrow\, S) \& (\neg A \,\Box\!\!\rightarrow\, S)$ to indicate that S would obtain whether or not A were performed.

---

[7] This way of speaking can lead to confusion about the objects of desires. When they are not careful, some philosophers seem to suggest that people desire propositions, or that desires are best understood as attitudes toward propositions. This is fine as a *façon de parler*, but wrong if taken literally. We desire propositions *with a specific truth-value*. The man who wants an extra $1,000 does not want the proposition that he wins $1,000; he wants the circumstances that would make this proposition true. Thus, a desire is always "directed toward" the potential states of affairs that can make a proposition true or false. It is not the proposition, but this state of affairs, that serves as the object of desire. The same remarks apply to beliefs. Even though it is sometimes convenient to pretend otherwise, beliefs are not ultimately about propositions, but about the possible states of affairs that determine their truth-values.

[8] Jeffrey (1965/1983, p. 84).

With all this logical apparatus at our disposal we can be clearer than Savage was about the internal structure of decision problems. Two broad structural features are fairly obvious. In a well-posed problem it ought to be a certainty that one and only one act will be performed and that exactly one state will obtain. **S** and **A** should thus be *partitions* of pairwise incompatible propositions exactly one of which is true as a matter of logic. Moreover, to ensure that there will never be any uncertainty about what outcome a given act/state pair will produce we must require A & S to entail $O[A, S]$ for every A and S. One of the keys to understanding decision making lies in getting clear about what the elements of these three partitions are, and about the logical interrelationships among them. We will take each of the three in turn, beginning with outcomes.

2.2 OUTCOMES

Intuitively, the proposition $O[A, S]$ is supposed to describe all of the desirable and undesirable things that the combination of A and S would bring about. The crucial word here is *all*. An outcome proposition in a well-posed decision problem should be detailed enough to supply a definite answer to *every* question the decision maker might care about in the context of her decision. It is important to keep this is mind because the single most common fallacy that people commit in the application of decision theory is that of underspecifying outcomes. The rule is very simple: O is underspecified any time there is a possible circumstance C such that the agent would prefer having O in the presence of C to having O in C's absence. Whenever there is such a circumstance O must be replaced by two more specific outcomes $O_1 = (O$ & $C)$ and $O_2 = (O$ & $\neg C)$. Failure to observe this simple rule has led even top-notch decision theorists to make elementary errors.

It is a straightforward consequence of this rule that the desirability of an outcome cannot be a function of either the state or the act that brings it about. In any well-posed decision problem the following two principles must hold for every $A, A^* \in$ **A** and $S, S^* \in$ **S**.[9]

---

[9] Cases of this sort are handled in a different way by proponents of so-called *state-dependent* utility theory. On state-dependent models, the desirability of a single outcome can vary across states of the world, so that having an extra $1,000 has one value given that the rate of inflation is low, and another given that the inflation rate is high. At a certain level there is no difference between this approach and the one advocated here. In most practical and theoretical contexts it makes no difference whether one portrays an agent as having conditional desires for unconditional outcomes or unconditional desires for outcomes with the relevant conditions packed into

State-Independence of Outcomes. If the agent does not find A's outcome under S exactly as desirable as its outcome under S^*, then $O[A, S]$ and $O[A, S^*]$ must be different propositions.

Act-Independence of Outcomes. If the agent does not find A's outcome under S exactly as desirable as A^*'s outcome under S, then $O[A, S]$ and $O[A^*, S]$ must be different propositions.

To see how the first might fail, suppose that you have a choice between two wagers on the rate of inflation for the rest of this year, with payments to be made at midnight on December 31. Your options are[10]

	The inflation rate will be 20% or less.	The inflation rate will be over 20%.
Bet on low inflation.	Win $1,000.	Lose $1,000.
Bet on high inflation.	Lose $1,000.	Win $1,000.

Obviously, the "Win $1,000" entry is worth more when it appears at the upper left than when it appears at the lower right. This violates our rule for individuating outcomes by making it appear as if altering the world's state can change the value of "the same" outcome. "Win $1,000" thus needs to be replaced by "Win $1,000 after inflation of less than 20%" at the upper left and by "Win $1,000 after inflation of more than 20%" at the lower right.

To see how outcomes might not have act-independent values consider the following passage from an article by Mark Machina, which is supposed to refute expected utility theory by offering the following putative counterexample to one of its central tenets:

Mom has a single indivisible item – a "treat" – to give to either daughter Abigail or son Benjamin. Assume that she is indifferent between Abigail getting the treat and Benjamin getting the treat, and strongly prefers each of these outcomes to the case where neither child gets it. However, in violation of the precepts of expected utility theory, Mom *strictly prefers* a [fair] coin flip to either of these sure outcomes.[11]

them. It is, however, one of the central theses of this book that the concept of a conditional desire is ambiguous between a causal and an evidential reading, and state-dependent decision theory, which starts from a largely unanalyzed notion of conditional preference, is ambiguous in exactly the same way. We cannot hope to remove this ambiguity, or even characterize it properly, unless we have access to outcomes that are unconditionally desired.

[10] I have based this example on a similar one found in Schervish et al. (1990).

[11] Machina (1991, p. 64). Machina is taking his example from Diamond (1967).

The decision Mom faces, according to Machina, is this (assuming that she is going to flip the coin anyhow):

	The coin lands heads.	The coin lands tails.
Give Abby the treat.	Abby gets the treat.	Abby gets the treat.
Give Ben the treat.	Ben gets the treat.	Ben gets the treat.
Allocate the treat via the coin toss.	Abby gets the treat.	Ben gets the treat.

If this is the right way to describe Mom's situation, then expected utility theory is indeed in hot water because Mom's preference for the third option is obviously rational, and the theory requires an agent who is indifferent between two outcomes to be indifferent between either of them and a lottery that offers a fifty-fifty chance of each. Fortunately, the theory's proponents need not be concerned; Machina has presented us with a red herring. Either Mom is not so obviously rational or the outcomes have been misdescribed. There are two possibilities to consider: Either Mom prefers the state of affairs in which a child receives the treat as a result of having won the coin toss to the state of affairs in which he or she receives it in the absence of a toss, or she does not. If the latter, then Mom is not rational since she strictly prefers tossing the coin to not tossing it even though she does not think the world will be a better place for its having been tossed. The reason why Mom seems rational, of course, is that we implicitly assume that she *wants* to allocate the coin via a fair method, and thus that the decision she faces is really the following:

	The coin lands heads.	The coin lands tails.
Give Abby the treat.	Abby gets the treat. via an unfair process.	Abby gets the treat. via an unfair process.
Give Ben the treat.	Ben gets the treat. via an unfair process.	Ben gets the treat. via an unfair process.
Allocate the treat via the coin toss.	Abby gets the treat. via a fair process.	Ben gets the treat. via a fair process.

What we have here, then, is not a counterexample to standard decision theory, but a case of underdescribed outcomes.[12]

Another kind of case in which the underspecification of outcomes misleads people into thinking that rational peoples' choices conflict with expected utility theory are those in which acts have "symbolic"

---

[12] This way of diagnosing the fallacy is not original with me. One can also find it in Broome (1990) and Pettit (1991), as well as many other places.

or "expressive" value. A nice illustration of this is provided by the "paradox of voter turnout." Suppose that you are trying to decide whether or not to vote in some large election. We imagine that there are nonnegligible costs to voting; for example, you might have to walk to the polling place in the rain or pay a poll tax. Here is one way of putting the problem you face (assuming that your candidate wins ties to keep the table small):

	Your candidate wins whether or not you vote.	Your candidate loses whether or not you vote.	Your candidate wins if you vote, but loses if you do not.
You vote.	Your candidate wins and you pay the costs of voting.	Your candidate loses and you pay the costs of voting.	Your candidate wins and you pay the costs of voting.
You do not vote.	Your candidate wins and you do not pay the costs of voting.	Your candidate loses and you do not pay the costs of voting.	Your candidate loses and you do not pay the costs of voting.

The paradox is that people turn out to vote at all. In any moderately large election, a million likely voters say, it is unreasonable to pay even a minor cost in the name of helping your candidate win because the probability that your vote will make the difference is so small as to be indistinguishable from 0 for *all* practical purposes.

The natural way of explaining away the paradox is to suppose that people vote not merely to help their candidate win, but because the act of voting itself has value for them. This may not be universally true; some people probably do vote for irrational reasons, but it is clearly the right explanation in many cases. Voting often has a *symbolic* or *expressive* value.[13] In casting a ballot one makes a kind of a

[13] The two philosophers who have written most eloquently about symbolic or expressive value are Robert Nozick (1993) and Elizabeth Anderson (1993). Nozick thinks of acts being performed on the basis of *principles* that situate them within some wider class of behavior. By adopting a principle one establishes a "symbolic connection" between the acts that fall under it, and this allows one to endorse all of them by performing any one. Anderson understands the expressive function of an act not by appealing to the other acts it "stands" for, but in terms of the way in which it conforms to social norms for expressing one's values about people and things. Despite their differences, both Nozick and Anderson will agree that a person can have a compelling reason to vote even when she is morally certain that doing so will make no difference to the outcome of the election because the act of voting might express something that is worth expressing no matter which candidate wins. Neither thinks, however, that this sort of symbolic or expressive value can be fully expressed in the outcomes of a decision table.

declaration to oneself, or others, that one is a certain sort of person. A vote may symbolize or express solidarity with a party or class, repugnance at some constellation of political ideas, and any number of other things. If this is the right explanation for many people, then the "paradox of voter turnout" is nothing more than a misdescribed decision problem. The "paradox" vanishes once each outcome in the top row of the matrix is augmented with a description of all the things that voting would express or symbolize and each outcome in the bottom row is augmented with a description of all the things that not voting would express or symbolize.

The most efficient way to avoid "outcome underspecification" within a propositional framework would be to require each outcome in a well-posed decision problem to *entail* the state and act propositions that bring it about. In other words, in addition to having A & S entail $O[A, S]$ one would also have $O[A, S]$ entail A & S, so that each outcome is the conjunction of the act and state that bring it about. This is what Jeffrey proposed. In his version of decision theory every finite decision matrix assumes the form

STATES

		S_1	S_2	S_3 ...	S_n
A	A_1	A_1 & S_1	A_1 & S_2	A_1 & S_3 ...	A_1 & S_n
C	A_2	A_2 & S_1	A_2 & S_2	A_2 & S_3 ...	A_2 & S_n
T	.	.	.	.	.
S	.	.	.	.	.
	.	.	.	.	.
	A_m	A_m & S_1	A_m & S_2	A_m & S_3 ...	A_m & S_n

OUTCOMES

Jeffrey's interpretation has some real advantages, but it turns out not to be an option within the confines of Savage's framework. The identification of $O[A, S]$ with A & S makes it impossible for the same outcome to appear at more than one place in a decision matrix, and in Savage's theory it is essential that this be able to happen. Since we will develop Savage's theory in the next chapter we need to allow a single outcome to be the result of more than one act/state pair. What we must do, therefore, is distinguish between the fine-grained *Jeffrey outcomes*, which are conjunctions of the form A & S for $A \in$ **A** and $S \in$ **S**, and coarse-grained *Savage outcomes*, which are disjunctions of Jeffrey outcomes that are equally desirable from the decision maker's point of view. The main thing to keep in mind about Savage outcomes is that they are *unalloyed goods* relative to the specification of acts and states in a decision problem. The agent should, in other words, always be indifferent between the truth of (O & A & S) and that of (O & A^* & S^*) for any $A, A^* \in \textbf{\textit{A}}$ and $S, S^* \in \textbf{\textit{S}}$. If this is not the case, then the problem is badly posed.

2.3 STATES

In some ways states are the easiest of the three components of decision problems to characterize. A state is a description of conditions in the world over which the decision maker exercises no direct control. That sounds simple enough, but the issue of what it means to "exercise no control" is a tricky one. We will see in subsequent chapters that Savage's framework for decision making makes sense only when the probabilities of the states in **S** are *independent* of the decision maker's desires and choice of actions. The problem is that this requirement of independence can be interpreted in at least two ways. It might mean either that the agent's desires and actions never give her *evidence* for thinking that one state rather than another will occur, or that they have no *causal influence* on the likelihoods of states. Let's leave these complications aside for a moment since they will be the focus of Chapters 4 and 5 and instead content ourselves with some general remarks about what sorts of information definitely must be included in the elements of **S** no matter how the independence requirement is cashed out.

Note first that that the more broadly we construe the concept of an action the less broadly we need construe the notion of a state; the more informative propositions in **A** are, the less informative those in **S** have to be to pick out the same set of outcomes. To see the contrast, consider walking to work versus *resolving* to walk to work. If we count the first of these as an act, then the states in our decision problem do not need to address the issue of whether the agent will be able to carry through on her resolution, but we will have to address this issue if the act is that of resolving to walk to work and the walking is a part of the outcome. My inclination is to adopt the second point of view and use the term *act* narrowly to denote *pure, present exercises of the will*. Many things we ordinarily call acts do not count as such under this reading. Walking to work, for example, is not really an act one can choose to perform because, unfortunately, one cannot simply will it to be the case that one's legs function properly, that one will not be shot by a madman on the way out the front door, that a great chasm will not open in the earth to prevent one from reaching one's destination, or even that one's "future self" will carry through on one's choice. We

57

ordinarily leave such contingencies unmentioned because they are so unlikely; in common speech the term "action" can be applied to any event that a person can make *highly probable* by an exercise of her will. We thus talk of people's choosing to walk to work, to have a piece of pie, to go to the dentist next week, and so on. This usage tends to mix acts, states, and outcomes in a misleading way, and thus it should be avoided. Strictly speaking what the agent has direct control over are the acts of her will; everything else is part of the state of the world. While I will sometimes speak in the colloquial way in what follows, talking of an agent's choosing to walk to the store, this is only a *façon de parler*. The act in question is always that of *willing* or *intending* oneself to behave in a certain way.

Given this way of looking at matters, the most obvious thing that should be included in state propositions is information about the past. Since no one can change the past by an act of will, any fact about history, *including facts about an agent's past acts*, that may make a difference to her happiness should be included in the elements of **S**. The decision maker is not, of course, required to know the past. Indeed, it is typically uncertainty about past events, even one's past acts, that makes decision making challenging.

Another, less obvious, kind of information that should be included in the states in **S** is that which pertains to *future choices*. Most of what we do consists in adopting *plans for future action* in which we resolve, in advance, to make certain choices should we face certain decision problems. Even behavior that plays out over a short time span is often best viewed as involving planning. When I decide to have celery as a snack, so as to maintain my diet, the act I choose is *not* that of eating celery. It is, rather, that of adopting a plan under which I to go to the refrigerator, open it, and then make a *further* decision to eat celery (rather than, say, the cherry pie I know to be in the refrigerator). The distinction here is between the adoption of a plan, which involves *resolving* to make certain future choices, and its execution, which requires that these choices be made. On the present construal the former is an act but the latter is not.

The reason for slicing things up this way is that people generally cannot, by a pure exercise of present will, force their future choices to conform to their current plans. There are a number of things that make such "predecision" impossible. First, it is always possible that unforeseen external events will prevent the choices envisioned in a plan from arising (unbeknown to me, my wife has thrown away the celery). Second, during the interval between the adoption of the plan and its execution the agent may become unwilling or unable to carry through on her earlier choices. This can happen either as the result of changes in opinions (while walking to the refrigerator I hear that celery causes cancer), or changes in intrinsic desires (upon seeing the celery I lose my appetite), or simple "weakness of the will" (seeing the pie next to the celery breaks my resolve and I eat pie "against my better judgment"). However it occurs, the point is that a decision maker cannot determine how her future choices will be made, or even what choices she will face, without the cooperation of at least some events in the world over which she has no direct control. This is why the *execution* of a plan is not, strictly speaking, an action.

In saying this I am diverging from the interpretation of acts that Savage offers in *The Foundations of Statistics*. In one of the book's most famous passages he illustrates his model of decision problems using the case of an omelet maker who must decide whether to break a sixth egg, which may be rotten, into a bowl containing five good eggs (perhaps spoiling them all); or to break the egg into a saucer (thus creating more dishes to wash); or just to make do with a five-egg omelet. He analyzes this case as follows:

> [My] formal description might seem inadequate in that it does not provide explicitly for the possibility that one decision may lead to another. Thus, if the omelet should be spoiled by breaking a rotten egg into it, new questions arise about what to substitute for breakfast.... Here I would say that the list of available acts envisioned [in the decision matrix] is inadequate for the interpretation that has just been put on the problem. Where the single act "break into bowl" now stands, there should be several, such as: "break into bowl and in case of disaster have toast," "break into bowl and in case of disaster take family to a neighboring restaurant for breakfast." ...
>
> What in the ordinary way of thinking might be regarded as a chain of decisions, one leading to the other in time, is in the formal description proposed here regarded as a single decision ... it is proposed that the choice of a policy or plan be regarded as a single decision.[14]

It is clear from this passage that Savage denies the "ordinary way of thinking" because he wants his theory to account for the fact that *outcomes* of present choices often depend on how future choices are made. He is surely correct in thinking that such dependencies exist, and that an adequate account of decision making must account for them. He attempted to do this by, in effect, portraying the decision maker as choosing among complete courses of future action that decide every issue over which he ever has any control. "The person," he writes, "decides 'now' once and for all; there is

---

[14] Savage (1954/1972, pp. 15–16).

nothing for him to wait for, because his decision provides for all contingencies."[15]

There is a way to account for the dependence of outcomes on future actions without going to such extremes. We can see how it works by asking why people make resolutions to perform future acts at all. Why do we even bother to make plans when we know that they can always be brought up for reconsideration and revision by our future selves? I think the answer should be obvious: In the typical case we adopt plans that require us to perform acts because doing so makes it *more likely* that we will perform those acts. I resolve to eat celery before I open the refrigerator door because this makes it (slightly) more likely that I will have the celery rather than the pie. Resolving to carry out a plan is a way of influencing the choices of our future selves.[16] When the matter is stated this way it is clear that the problem of accounting for the dependence of outcomes on future acts is not best handled by incorporating the latter into **A** as Savage does, but by incorporating them into the states of **S** since these are supposed to serve as the locus of uncertainty in the model. A person deciding whether to adopt a plan that would require her to choose to perform some act A at future time t should be portrayed as facing a decision in

---

[15] Savage (1954/1972, p. 17).

[16] I do not pretend to know what psychological mechanisms are involved in this process. Part of the story is surely provided by Michael Bratman's influential analysis of intentions and plans found in his (1987). On Bratman's view the adoption of a plan involves a defeasible commitment not to reopen deliberations on advisability of executing the acts that the plan prescribes. It is plausible to think that having such commitments can make it less likely that one will revisit questions one has already decided; one sometimes carries out the plans of one's past self in the same unquestioning way that a soldier near the end of boot camp carries out the commands of his drill sergeant. Thus, resolving to do A can make it more likely that one will do A simply by making it less likely that one will have a chance to change one's mind. Another part of the story, not in conflict with the first, is that a person who plans to do A may thereby give her future self a *reason* to do A; that is, in the future she might take the fact that A accords with her past plan as a consideration in favor of doing it, and this make it more likely that one does it. One of the easiest, and I suspect most common, ways for this to happen has to do with the fact that adopting a plan to do A can give one's future self *evidence* for thinking that one would decide to do A were one to think about the matter in detail. This, of course, gives one's future self a reason *not* to think about the matter in detail. For example, a traveler who planned her route carefully with map in hand might, upon coming to a fork in the road, take the left branch because she knows that this seemed the best option to her when she had time to think about the matter in detail. This gives her evidence for thinking that she would decide the same way now, which is a reason for taking the left branch now. Among the authors who see past plans as providing *nonevidential* reasons for future actions are David Velleman (1989) and Edward McClennen (1990). Whatever the mechanisms involved here, the general point is that people make plans with the intention of influencing the likelihood that they will make certain future decisions and perform certain acts.

which every state of the world specifies that one of the following four contingencies holds:

C_1: I will choose to do A at t whether or not I resolve to do so now.

C_2: I will choose to do A at t if I resolve to do so now, but not otherwise.

C_3: I will not choose to do A at t if I resolve to do so now, but will otherwise.

C_4: I will not choose to do A at t whether or not I resolve to do so now.

If the agent presently regards the consequences of choosing A to be better than those of not choosing it, then it will make sense for her to plan to do A insofar as she thinks that C_2 is more likely than C_3. The rational decision maker should thus treat her future choices and acts as being on a par with states of the world since they are, strictly speaking, events that presently lie beyond her direct control. To treat them in any other way is to misunderstand the distinction between events that are probable consequences of one's present decisions *given the state of the world* and events that one can bring about by an exercise of one's present will *no matter what state the world is in*. Only the latter should be counted as acts in the strict sense of the term.

Aside from past facts and facts about future choices, there will be many other sorts of facts that the states in **S** should specify. It would not be worthwhile to try to catalogue all of them here. In most cases an intuitive understanding of states as "descriptions of aspects of the world that lie outside the decision maker's control" will suffice. The real problem cases are the ones we will consider in later chapters. For now, let us take the notion of a state as sufficiently well understood and move on to acts.

2.4 ACTIONS

One of the principal differences between the models of decision problems offered by Savage and Jeffrey lies in their disparate treatments of the concept of an *action*. Jeffrey identifies acts with propositions that a decision maker can make true or false as she pleases. In most cases this coheres well with our intuitive notion of an act as a "piece of behavior directly controlled by an agent's will," but we shall see below that Jeffrey's criterion is slightly too liberal since a proposition can satisfy it without describing intentional behavior of any kind. The entities that Savage calls actions have an even less immediate connection to willed behavior. In fact, as we will see, if Savage's theory is

61

to be plausible at all, then his "acts" must be understood simply as objects of instrumental desire whose realization may or may not lie within the agent's control. Let's start by discussing Savage's model of actions.

Savage portrays acts as abstract functions that assign outcomes to states of the world. In keeping with the propositional interpretation being advocated here, I am going to replace these functions by the propositions that describe them. The challenge is to find the right sort of proposition to do the job. One natural way to proceed would be to notice that since $A \ \& \ S$ entails $O[A, S]$ for every S, it will always be true that each A in **A** entails a conjunction of material conditionals of the form

$$(*) \quad \&_S (S \supset O[A, S])$$

where S ranges over **S** and $O[A, S] \in \textbf{\textit{O}}$. (*) suggests a functional dependence of outcomes on states that is in keeping with Savage's view of acts. When a decision maker "performs" the act associated with the function that takes S to $O[A, S]$ she will automatically make the proposition $\&_S (S \supset O[A, S])$ true. One might thus be led to wonder whether acts can simply be *identified* with propositions of form (*).

This is what Jeffrey proposed in "Probable Knowledge." (He has since altered his views). "The situation in which a gambler takes himself to be gambling on S with prize W and loss L," he wrote, "is one in which he believes the proposition $G = (S \ \& \ W) \vee (\neg S \ \& \ L)$."[17] Since G is logically equivalent to $(S \supset W) \ \& \ (\neg S \supset L)$, Jeffrey can be read as claiming that the performance of an action always involves making a proposition of form (*) true. This cannot be correct.[18] While Jeffrey was quite right to think that an agent who chooses to stake the difference between W and L on the occurrence of S is making G true, he was wrong to imply that this is all she makes true. If S happens to be false, then the agent can make G true merely by making its second conjunct $\neg S \supset L$ true. But if she can do this, then she can make *any* proposition of the form $(S \supset X) \ \& \ (\neg S \supset L)$ true. To see why this is problematic, suppose that, for a cash payment of $10 today, I agree to make the following statement true:

G: Either it rains tomorrow and I give you $100, or it does not rain tomorrow and I give you $5.

---

[17] Jeffrey (1968, p. 167). I have changed the symbols slightly in this quotation to bring it into line with the notation being used here.
[18] For additional criticisms of Jeffrey's view of acts see Spohn (1977).

Once the deal is done you discover that I am going to be unable to pay up if it rains because I have only $50 to my name (your $10 included). Fortunately for me, but unbeknown to both of us now, the point is moot since tomorrow is going to be a sunny day. Now, have I really done what I said I would do if I am prepared to pay you $5 if it does not rain but not prepared, or even able, to pay you $100 if it does rain? Obviously not! For me to hold up my end of the bargain it has to be the case that I would pay you were it to rain. Notice, however, that all it takes to make the proposition G true is my willingness and ability to pay you $5. This is also sufficient to guarantee the truth of the following proposition:

H: Either it rains tomorrow and I give you $50, or it does not rain tomorrow and I give you $5.

Indeed, since it is not going to rain tomorrow it follows that by paying you $5 I do everything I need to do to guarantee the truth of *every* proposition of the form:

Either it rains tomorrow and I give you x, or it does not rain tomorrow and I give you $5.

This shows that, on Jeffrey's interpretation, there is no distinction to be made (in the event of no rain) between my act of making G true and also making it true that I would pay up if it were to rain, and my act of making G true and also making it true that I would not pay up if it were to rain.

This is the wrong result. It is crucial to our intuitive understanding of actions, and to Savage's picture of them, that two acts should count as distinct when they have incompatible outcomes in any state of the world *whether or not that state happens to obtain*. Even if it fails to rain tomorrow, if I would not pay you $100 if did, then I have performed a different act from the one I would perform if I would make such a payment. In the end, the problem with identifying acts with propositions of the form (*) is that it leaves us with too coarse a notion of an act.

To obtain a propositional expression of the relationship between states and outcomes that Savage envisions, and that our common conception of action requires, we need to make sure that two act propositions are contraries whenever they have contrary outcomes in any state of the world, whether this state is actual or not. One way to do this is by identifying an act with a proposition of the form

$$(**) \quad A = \&_s (S \Rightarrow O[A, S])$$

63

where $\Rightarrow$ is a kind of *conditional* that is stronger than the ordinary material conditional. We will discuss conditionals at some length in Chapter 5, but for now it suffices to know that a conditional is any binary propositional connective that satisfies the following axioms:[19]

Cond₁ (*Entailment*). If χ entails φ, then $\chi \Rightarrow \varphi$ is a logical truth.

Cond₂ (*Centering*). χ & $(\chi \Rightarrow \varphi)$ is logically equivalent to χ & φ.

Cond₃ (*Weakening the Consequent*). $\chi \Rightarrow \varphi$ entails $\chi \Rightarrow (\varphi \vee \psi)$.

Cond₄ (*Conditional Conjunction*). $(\chi \Rightarrow \varphi)$ & $(\chi \Rightarrow \psi)$ entails $\chi \Rightarrow (\varphi \& \psi)$.

Cond₅ (*Dilemma*). $(\chi \Rightarrow \varphi)$ & $(\psi \Rightarrow \varphi)$ entails $(\chi \vee \psi) \Rightarrow \varphi$.

Cond₆ (*Weak Strengthening of the Antecedent*). $(\chi \Rightarrow \varphi)$ & $\neg(\chi \Rightarrow \neg\psi)$ entails $(\chi \& \psi) \Rightarrow \varphi$.

Cond₇ (*Reductio*). $\chi \Rightarrow (\varphi \& \neg\varphi)$ entails $\psi \Rightarrow \neg\chi$.

Cond₈ (*Conditional Equivalence*). If $(\chi \Rightarrow \varphi)$ & $(\varphi \Rightarrow \chi)$, then $\chi \Rightarrow \psi$ if and only if $\varphi \Rightarrow \psi$.

Among the operators that meet these requirements are the material conditional, the subjunctive conditional, "If — were the case then ___ would also be the case," and perhaps the indicative conditional "If ---- is the case then ____ is also the case" (although there are many who doubt that the indicative conditional expresses a proposition).

If the identification of **A** with the family of all statements of form (**) is to make sense, then these statements must form a partition. This requires $\Rightarrow$ to have three properties that are not shared by all conditionals. First, if (**) is to avoid the problems associated with (*) then the elements of **A** have to be contraries. In the presence of the other axioms, we can guarantee this by having $\Rightarrow$ satisfy

Conditional Contradiction. If φ and φ^* are contraries and χ is not a contradiction, then $\chi \Rightarrow \varphi$ and $\chi \Rightarrow \varphi^*$ are also contraries.

Second, since *every* function from states to outcomes is supposed to be a possible act in Savage's model we also need to suppose that "$\Rightarrow$" obeys the principle of

Harmony. If χ and χ^* are contraries, and neither $\chi \Rightarrow \varphi$ nor $\chi^* \Rightarrow \varphi^*$ is a logical falsehood, then $(\chi \Rightarrow \varphi)$ & $(\chi^* \Rightarrow \varphi^*)$ is not a logical falsehood.

[19] Discounting for minor stylistic differences, this is a fairly standard axiomatization of "centered conditionals." It is equivalent to those offered in Gardenfors (1988, pp. 148–52) and in Lewis (1973, p. 132).

Given the other axioms, this ensures that *every* conjunction of the form (**) will be logically consistent. Finally, to ensure that the elements of **A** are collectively exhaustive we need to add

Conditional Excluded Middle. $(\chi \Rightarrow \varphi) \vee (\chi \Rightarrow \neg\varphi)$ is a logical truth.

Because **O** and **S** are partitions one can use this principle to show that **A** must be a partition too.

The moral here is that if there is a propositional operation $\Rightarrow$ that obeys Cond_1–Cond_8, Conditional Contradiction, Harmony, and Conditional Excluded Middle, then it makes sense to identify **A** with the set of all propositions of the form $A = \&_S (S \Rightarrow O[A, S])$ where $O[A, S]$ may be any outcome in **O**. The question is whether such operators exist. It is not clear that they do. While the subjunctive conditional does satisfy Conditional Contradiction and Harmony, it violates Conditional Excluded Middle because, as David Lewis has convincingly argued, some counterfactuals are neither true nor false, such as "Had Bizet and Verdi been compatriots they would have been Italian."[20] The indicative conditional does seem to satisfy all three requirements, but, as we will see in Chapter 6, there are many philosophers who doubt that indicative statements in the conditional mood even express propositions. These issues, however, are best left for later discussions (where we will see how they can be finessed). The important point at the moment is that we cannot adequately formulate Savage's theory without assuming the existence of a conditional $\Rightarrow$ that obeys the three principles in question in cases where its antecedent is a state in **S** and its consequent is an outcome in **O**. I will call such an operation a *Savage conditional*. Given the existence of a Savage conditional it makes sense to define **A** as the collection of conjunctions of the form $A = \&_S (S \Rightarrow O[A, S])$ where S ranges over states in **S** and $O[A, S]$ may be any outcome in **O**.

Savage's decision theory (to be described in the next chapter) assumes that **A** has a great deal of internal structure. First, it requires the existence of "constant" acts that produce the very same outcome in every state of the world. For each outcome O there is supposed to be an act $A_O \in \mathbf{A}$ such that $O = O[A_O, S]$ for each S. A person who "performed" A_O would thereby insulate himself from the effects of *all* external contingencies so that, from his point of view, it would no longer matter what the world was like. He could be certain that his lot would be O no matter what state the world was in. We will see that

[20] Lewis (1973, p. 80).

65

the assumption of constant acts is one of the major shortcomings of Savage's approach.

The theory also assumes that **A** must be closed under a certain kind of "mixing."[21] To define mixing in this context we must introduce the notion of a *conditional* action. Suppose that E is a disjunction of states, hereafter an *event*. For any act $A = \&_S (S \Rightarrow O[A, S])$, define *A-conditional-on-E* as the proposition

$$A_E = \&_S ((S \& E) \Rightarrow O[A,S]).$$

Since $(S \& E) \Rightarrow O[A, S]$ is true whenever its antecedent is a contradiction (by $Cond_1$), this act is "performed" whenever an unconditional act whose outcomes coincide with A's on E is made true. A_E thus specifies a "partial" function that assigns A's outcomes to states that entail E, but which assigns no outcome to states that are incompatible with E.[22] Since **A** is a partition, and since $A_E \& B$ is contradictory unless $O[A, S] = O[B, S]$ whenever S entails E, it is always possible to write A_E as the disjunction of unconditional actions whose outcomes coincide with A's on E, so that $A_E = \vee_{B \in \mathbf{A}}(A_E \& B)$ $= \vee\{B \in \mathbf{A}: O[A, S] = O[B, S], S \text{ entails } E\}$.

Two most important facts about conditional acts are

Decomposition. $A = (A_E \& A_{\neg E})$.

Mixing. $(A_E \& B_{\neg E}) \in \mathbf{A}$ whenever $A, B \in \mathbf{A}$.

(I invite the reader to derive these as consequences from the definition of a conditional act and the properties of Savage conditionals.) The "mixed" act $(A_E \& B_{\neg E})$ describes a situation in which the decision maker is certain to secure A's outcome when E is true and to secure B's outcome when E is false. An example might be helpful here. Suppose you are sending a friend out to buy your lunch at a local restaurant. Your options might be

	Chili is the special.	Pizza is the special.
A = Order chili.	Eat cheap chili.	Eat expensive chili.
B = Order pizza.	Eat expensive pizza.	Eat cheap pizza.
A_E & $B_{\neg E}$ = Get the special.	Eat cheap chili.	Eat cheap pizza.

---

[21] In fact, Savage assumes something stronger than this, namely, that *every* mathematically possible function from states to outcomes will be in **A**. Strictly speaking, however, his formalism requires only closure under mixing.

[22] It is consistent with the laws governing Savage conditionals that A_E is equivalent to $E \Rightarrow A$, but this is not mandatory.

If you are concerned about expenses and have no clear preference between chili and pizza, then the mixed act is the way to go. The point to keep in mind here is that mixed acts of this sort are *always* available in Savage's framework. This is one of its great theoretical advantages and, as we shall see, one of its great practical weaknesses.

Given that **A** has all this structure in Savage's system it is quite misleading to refer to its elements as "acts" since many of them do not correspond in any reasonable sense to things that an agent can bring about by a mere exercise of her will. This is especially clear in the case of constant acts. No human has the power to perform such "acts."[23] A being who could carry them out would be omnipotent since she could, in effect, choose outcomes directly. Even if such a being were to exist, the point of *expected* utility theory would be lost on her. She would need only one decision rule, and it would be simplicity itself: Maximize nonexpected utility by performing the constant act associated with the very best outcome in **O** if there is a best outcome. (If there is no best outcome in **O** then the agent has no rational choice.) So, insofar as we are dealing with agents who *need* a theory of expected utility – those who need to consult their beliefs about the state of the world in order to make wise choices – we must reject the idea that elements of Savage's **A** correspond to things that the decision maker can "do." It is better to think of them as *states of instrumental value*, which sometimes will, but more commonly will not, lie within a decision maker's direct control.

2.5 DECISION PROBLEMS

At this point it will be useful for us to stand back a bit and formulate an abstract and general characterization of decision problems that can incorporate both Savage's model, with its highly developed set of "acts," and Jeffrey's less complicated model. We can start by thinking of the partitions **A**, **S**, and **O** as embedded within a larger set of propositions Ω that has the structure of a Boolean σ-algebra. Ω can be defined as the smallest collection of propositions that includes the partitions **O**, **A**, and **S**; is closed under negation, so that $\neg X \in \Omega$ whenever $X \in \Omega$; and is closed under countable disjunction, so that $\vee_j X_j = (X_1 \vee X_2 \vee X_3 \vee \ldots) \in \Omega$ whenever $X_1, X_2, X_3, \ldots \in \Omega$. Ω will

---

[23] The only plausible candidate for a constant act I can think of is one in which a person, *who does not care at all how the future plays out after he is dead*, chooses to commit suicide. Even here, though, the person would have to consider the possibility that his attempt at self-annihilation will fail for some reason. So, even this is not a constant act.

be called the decision problem's *base algebra*, and the problem itself will be identified with the structure $D = (\Omega, O, S, A)$.

To take into account both Savage's and Jeffrey's views of outcomes we will require each act/state conjunction $A \& S$ to entail exactly one outcome, which means that O will be a coarsening of the partition of Jeffrey outcomes $W = \{A \& S: A \in A, S \in S\}$. I want readers to think of W as containing the "possible worlds" that are relevant to the decision problem. More precisely, the act/state conjunctions in W will be the *atomic* components of Ω in the sense that every proposition in the algebra will be uniquely expressible as a disjunction of them. Insofar as elements of Ω are concerned, the statements in W provide maximally specific, consistent descriptions of the possibilities in the sense that the truth-value of every proposition in Ω is decided once the true member of W is specified. Notice, though, that the "worlds" in W need not be *the* possible worlds, which are the atoms of the algebra of *all* propositions. Rather, they are the possible worlds insofar as the decision problem D is concerned.

The base algebra will contain many propositions other than those in A, S, and O. Some of the most important are what might be called *coarse-grained* actions, states, and outcomes. A coarse act is an Ω proposition X such that each element of A entails either X or $\neg X$; that is, X is a disjunction of elements of A. (As explained previously, conditional acts can be expressed this way.) Likewise, a coarse state, or *event*, is an Ω proposition E that can be expressed as a disjunction of elements of S. A coarse outcome is a disjunction of the elements of O. Each of these three kinds of propositions will appear in its own subalgebra of Ω. For future reference let's define

$\Omega(S)$, the subalgebra of *events*, as the set of all Ω propositions that are expressible as disjunctions of elements of S

$\Omega(A)$, the subalgebra of *coarse acts*, as the set of all Ω propositions that are expressible as disjunctions of the elements of A

$\Omega(O)$, the subalgebra of *coarse outcomes*, as the set of all Ω propositions that are expressible as disjunctions of the elements of O

The propositions in these three subalgebras turn out to play somewhat different roles in the decision theories that will be developed here. The one point of agreement is that the outcomes in O serve as the objects of the decision maker's basic desires, and that the objects in A are among the objects of her instrumental desires.

Decision theorists describe these desires, both basic and instrumental, ordinally in terms of a *preference ranking*. This is a pair of binary relations, $>$ and $\geq$, of strict and weak preference that characterize the comparative strengths of the decision maker's desires for the truth of propositions in Ω. $X > Y$ (or $X \geq Y$) holds exactly if she finds the prospect of X's truth strictly more desirable than (at least as desirable as) the prospect of Y's. When the agent is indifferent between the two prospects, so that both $X \geq Y$ and $Y \geq X$ hold, I will write $Y \approx X$. I am employing two primitive preference relations rather than the usual one because I do not want to assume from the outset that the agent's desires *completely order* Ω, $\Omega(\boldsymbol{A})$, or $\Omega(\boldsymbol{O})$. Many decision theorists do make such assumptions, and this allows them to take $\geq$ alone as primitive and to define $X > Y$ as not-$(Y \geq X)$. Despite its seeming advantages in terms of parsimony, I think this is an unwise course of action because it precludes the possibility that both $Y \geq X$ and $X > Y$ might fail to obtain for certain X, Y pairs. If such cases are to be precluded, this should be stated in the theory's axioms rather than being made part of the notion of a decision problem itself. Part of the reason for this is methodological: Even if we think the completeness of preferences is a requirement of rationality we still want to be able to describe irrational preferences. More importantly, though, many decision theorists reject completeness on the grounds that there is really nothing irrational about having desires that are vague or indeterminate.[24] Savage himself seemed sensitive to this point at one stage, writing that

There is some temptation to explore the possibilities of analyzing preferences among acts as a *partial ordering* ... [by] admitting that some pairs of acts are incomparable. This would seem to give some expression to introspective feelings of indecision, which we may be reluctant to identify with indifference.[25]

He immediately conjectures, however, that this would be a "blind alley," while granting that "only an enthusiastic exploration could shed real light on the question." We will do some of this exploring in the next chapter after we have presented Savage's theory.

It is useful to understand differences in decision theorists' views about the nature of decision problems as being either disagreements over what the components of the partitions $\boldsymbol{A}$, $\boldsymbol{S}$, and $\boldsymbol{O}$ should look like, or disputes about the propositions over which preferences and subjective probabilities may be properly defined. As we have seen, Jeffrey maintains that $\boldsymbol{A}$ can be any partition of propositions whose

[24] See, for example, Kaplan (1996, pp. 22–31).
[25] Savage (1954/1972, p. 21).

truth-values the decision maker can directly control, and he identifies **O** with the set of all act/state conjunctions. Savage, in contrast, needs his **A** to be a family of propositions of the form $A = \&_S (S \Rightarrow O[A, S])$, where "$\Rightarrow$" is a Savage conditional, which contains all constant acts and is closed under mixing.

Savage and Jeffrey also differ over the issue of what propositions should serve as the basic objects of desire and belief. Jeffrey takes the view that *any* proposition in Ω can figure into an agent's preference ranking or her subjective probabilities, whereas Savage thinks that only outcomes and acts can be objects of desire and that only states can be objects of belief. A third view that we will consider, which is due to Duncan Luce and David Krantz,[26] agrees with Savage about the last point, but takes a step in Jeffrey's direction by supposing that the agent's preference ranking will be defined over conditional acts of the form A_E. We shall consider the details of these models in the next two chapters. For the moment the important thing is that all of them can be understood as pertaining to agents who are facing decision problems of the form $\mathbf{D} = (\Omega, \mathbf{O}, \mathbf{S}, \mathbf{A})$ and who have preference rankings defined over some subset of Ω.

2.6 "SMALL WORLDS" AND THE "GRAND WORLD"

To apply the preceding model to the sorts of choices that human beings actually face we must recognize that there will usually be a vast number of decision problems that an agent can consider in a given context. Take our woman who is deciding whether to walk or drive to work on a cloudy day. We portrayed her as making the decision of Table 1. If there is an umbrella in the house, however, she could just as well be portrayed as facing the decision of Table 2. Alternatively, we could picture her taking the possibility of heavy rain into account by using the matrix of Table 3. There are clearly many other decisions that our agent might consider in this situation, and in the course of deliberating she may entertain a number of them.

The process of deliberation is often one of *refinement*. A decision maker begins by considering one family of possible actions, states of the world, and outcomes and moves on to consider others that provide a fuller and more realistic picture of her situation. When it comes to making a definitive choice she will settle on a *single* decision problem by focusing on a fixed repertoire of acts and defining outcomes in

---

[26] Luce and Krantz (1971).

	It rains.	It does not rain.
I walk to work.	I walk to work in the rain.	I walk to work, but not in the rain.
I drive to work.	I drive to work in the rain.	I drive to work, but not in the rain.

Table 1

	It rains.	It does not rain.
I walk to work and leave the umbrella at home.	I walk to work in the rain and do not have to carry an umbrella.	I walk to work but not in the rain, and I do not carry an umbrella.
I walk to work and take the umbrella.	I walk to work in the rain and must carry an umbrella.	I walk to work but not in the rain, and I must carry an umbrella.
I drive to work.	I drive to work in the rain.	I drive to work, but not in the rain.

Table 2

	It rains and traffic is heavy.	It rains and traffic is light.	It does not rain and traffic is heavy.	It does not rain and traffic is light.
I walk to work.	I walk to work in the rain.	I walk to work in the rain.	I walk to work but not in the rain.	I walk to work but not in the rain.
I drive to work	I drive to work in heavy traffic and in the rain.	I drive to work in light traffic and in the rain.	I drive to work in heavy traffic but not in the rain.	I drive to work in light traffic but not in the rain.

Table 3

terms of a specific partition of states. The character of the decision problem that she ultimately sets for herself will depend upon how thoroughly she reflects on her predicament. A person who makes the decision of the first matrix is not thinking very much at all: She is overlooking obvious options (like taking an umbrella) and is ignoring contingencies that could affect the outcomes of her choice (like the fact that she may get stuck in traffic). If she reflected on her situation

71

more deeply she would surely set herself a decision problem that involves a more detailed description of both her options and the world's state. A certain amount of refining is thus indispensable to making a rational choice. Choosing is really a two-stage process in which the agent first refines her view of the decision situation by thinking more carefully about her options and the world's state until she settles on the "right" problem to solve and then endeavors to select the best available course of action by reflecting on her beliefs and desires in the context of this problem.

Decision theorists have concentrated almost exclusively on the second stage of this process. Once the decision problem is in place they try to explain what makes the choice of an action rational or irrational. The initial stage is equally important, however, and any complete account of decision making must have something to say about it. First, there is the formal question of how to model the refinement process. This turns out to be fairly easy within the framework we have constructed. Say that one decision $\boldsymbol{D}^+ = (\Omega^+, \boldsymbol{O}^+, \boldsymbol{S}^+, \boldsymbol{A}^+)$ is a *refinement* of another $\boldsymbol{D} = (\Omega, \boldsymbol{O}, \boldsymbol{S}, \boldsymbol{A})$ just in case $\boldsymbol{O}^+$ is a refinement of $\boldsymbol{O}$, $\boldsymbol{S}^+$ is a refinement of $\boldsymbol{S}$, and $\boldsymbol{A}^+$ is a refinement of $\boldsymbol{A}$. (It follows that Ω must be a subalgebra of Ω^+.) When an agent clarifies some question for herself she moves from describing her situation by using some coarser decision problem $\boldsymbol{D}$ to describing it in terms of some finer one $\boldsymbol{D}^+$.

Now there would be no point to this process of refinement unless it sometimes led people to change their minds, which it often does. Decision makers' beliefs and desires about the states, acts, and outcomes in less refined problems often change when they consider their situation in greater detail. We generally think that more considered judgments are better, all else being equal, and that a rational decision maker has an obligation to consider her situation in "sufficient" detail before coming to some final view about what to want or do. One challenge is to say precisely how much refining a person needs to do before settling on the right problem to solve. Some refinements are clearly useless because they introduce considerations that are irrelevant to the decision maker's concerns. If a woman who is deciding whether to walk or drive to work does not care whether she starts off by stepping with her left foot or her right, it would be pointless for her to divide the act "I walk to work" into "I walk to work and begin with my left foot" and "I walk to work and begin with my right foot." Likewise, if it does not matter to her whether the temperature outside is 72°F or 73°F, then the states in her problem need not discriminate these contingencies. Even when such possibilities are excluded, however, the number of things that could be relevant to the decision maker's concerns is potentially infinite. Indeed, there is a sense in which the space of possible acts is *always* infinite because, for any n, the agent always has the option of resolving to spend the rest of her days trying to compute π to the nth decimal place. (I invite the reader to think up a similarly goofy reason for thinking the space of relevant states is always infinite.)

It is possible to theorize about the decision problem that the agent would come to if she could somehow take all her possible options into account and could resolve the partition of states to the highest level of pertinent detail. Savage called this her *grand-world decision*. The defining feature of a grand-world decision is that *any* consideration that could be relevant to the agent's desires is taken into account in the specification of acts and states. In the framework being developed here, the test is this:

$\boldsymbol{D}^{\text{G}} = (\Omega^{\text{G}}, \boldsymbol{O}^{\text{G}}, \boldsymbol{S}^{\text{G}}, \boldsymbol{A}^{\text{G}})$ is the *grand-world decision problem* that an agent faces if and only if there is no proposition X, whether in Ω or not, such that she strictly prefers $(O \ \& \ X)$ to $(O \ \& \ \neg X)$ for some outcome $O \in \boldsymbol{O}^{\text{G}}$.[27]

In other words, $\boldsymbol{D}^{\text{G}}$ is the decision problem whose outcomes function as unalloyed goods relative not only to the propositions in Ω, but to all the propositions that there are. When a decision problem fails this test it is a *small-world* decision.[28] Every small-world decision $\boldsymbol{D}$ is a coarsening of the grand-world problem $\boldsymbol{D}^{\text{G}}$, and there is always a sequence of refinements $\boldsymbol{D}, \boldsymbol{D}_1, \boldsymbol{D}_2, \ldots, \boldsymbol{D}^{\text{G}}$ that begins with $\boldsymbol{D}$ and ends with $\boldsymbol{D}^{\text{G}}$.

It should be clear no actual human agent is ever capable of contemplating even a small part of the grand-world decision she faces. To do so, she would have to consider not only the choices she must now make, but all the choices she might make over the whole course of her life. Even this, which is only the tip of the iceberg, lies far, far beyond human capacities. Deliberation, if it is to be any use at all, must stop well short of a grand-world problem. What needs to be explained, then, is how an agent's decisions about what to want and do can be rational given that they are, from a certain perspective, always *massively* underconsidered.

In answering, let's note first that the point at which an agent should stop refining cannot itself be a matter of rational decision making, for

[27] $\boldsymbol{D}^{\text{G}}$ is not, strictly speaking, unique; it is only unique up to *irrelevant* refinements.
[28] This does not agree precisely with Savage's use of the term, but it is close.

this would invite an infinite regress. Some have taken this as proof that rational choice is impossible. A particularly clear statement of the kind of reasoning that leads to this conclusion has been presented by Michael Resnick (who does not endorse the reasoning):

Whenever we apply decision theory we must make some choices: At least, we must pick the acts, states, and outcomes to be used in our problem specification. But if we use decision theory to make those choices, we must make yet another set of choices. This does not show that it is impossible to apply decision theory. But it does show that . . . any application of the theory must be based ultimately on choices that are made without its benefit . . . [These initial] decisions are not rational, and because all decisions depend ultimately on [them], no decisions are rational.[29]

The fallacy here is in the final inference from "the initial decisions are not rational" to "no decision is rational." We can grant that the decision to stop deliberations at some small-world problem D is often made on the basis of factors that are, to some extent, arational, but this arationality does not infect the decision made in D. What determines whether the agent's decision in D is rational are (a) the extent to which her decisions *in* the context determined by D obey the laws of rationality and (b) the extent to which she is *justified* in thinking that her evaluations of the merits of the small-world actions in D agree with the evaluations that she would give those same acts if she viewed them from the perspective of the grand-world problem D^G. This last point will take some explaining.

The beliefs and desires of an agent who faces a small-world decision problem have a different character from those of a grand-world decision maker. The latter attitudes are *fully considered*, whereas the former are not. A fully considered belief is one that a person holds after taking *all* the evidence at her disposal into account and drawing inferences from this evidence using the best principles of inductive and deductive logic. A fully considered desire is one that is based on complete and explicit reflection on all the contingencies that might affect the values of the prospects the person is considering. Fully considered beliefs and desires, in short, are the ones that an individual would hold if she were able to deliberate, in an entirely rational manner, on all issues relevant to her concerns (without additional input from the world) until further deliberation was superfluous; they are the "fixed points" of rational deliberations. The decision to perform a certain A, rather than any of the alternatives, is fully (less than

[29] Resnick (1978, p. 11).

fully) considered when it is based on fully (less than fully) considered beliefs and desires.

Of course, we flesh-and-blood human beings have neither the time nor the cognitive resources to reason ourselves anywhere near a state of reflective equilibrium. We therefore never really know what our fully considered attitudes are, and this prevents us from making the sorts of choices that grand-world decision problems require. Despite this, I want to claim that every time we make a decision, no matter how ill considered, we tacitly *commit* ourselves to holding certain views about the nature of our fully considered beliefs and desires and about the evaluative judgments they would sanction. Specifically, when we choose to perform a small-world act A in D on the basis of less than fully considered beliefs and desires we thereby commit ourselves to the view that our fully considered beliefs and desires would sanction the choice of A from among the alternatives listed in D. We cannot, in other words, choose to do A without committing ourselves to the claim that we would not rank some other act in D above A were we to deliberate on the matter, in an entirely rational way, until all further deliberation became moot.

In thinking about this point it is instructive to consider an analogy with "indefeasibility" theories of epistemic justification. A subject is said to be *indefeasibly* justified in (fully) believing a proposition P just in case she is justified in believing P and there is no further *true* proposition Q such that she would no longer be justified in believing P were she to learn Q.[30] This is a kind of *stability* condition. It says that the subject's justification should be sufficiently comprehensive that it cannot be *undermined* by any new factual information. Indefeasible justification was once touted as the elusive "fourth condition" of knowledge, but support for it waned once it became widely appreciated that a person can know P even if there is *misleading* evidence she does not possess that would undermine her justification for it. To account for the possibility of misleading evidence, the simple indefeasibility analysis had to be replaced by a weaker account that only required a knower's justification for P to be *ultimately undefeated* in the sense that any misleading defeater Q could be associated with another true proposition Q^* that would reverse Q's misleading effect so that the agent would still be justified in believing P were she to learn both Q and Q^*.

It is important to emphasize that it is no part of the indefeasibility analysis that the knower must *know* that her justification is ultimately

---

[30] For a recent defense of the indefeasibility analysis see Moser (1989).

indefeasible. Such a requirement would make knowledge impossible. As Gilbert Harman observed, however, it does seem that a knower is at least tacitly committed to the *belief* that her justification is ultimately undefeated; she is committed to thinking that there is no true, nonmisleading evidence that would undermine her justification for P.[31] A person who adopted the reverse attitude would (fully) believe P and yet think that there exists true nonmisleading evidence, unbeknown to her, that would undermine her justification for P. Since the latter thought itself undermines her justification for P, these two attitudes cannot be simultaneously held by a rational agent.

What I am proposing is that we treat the possible results of deliberation in decision making in much the same way that proponents of the sophisticated indefeasibility analysis treat the potential acquisition of evidence. The acquisition of evidence and deliberation are both learning processes – in the first we learn about ourselves and the world, while in the second we learn, in some sense, what *we think* about ourselves and the world. The line I want to push, in analogy with Harman's, is that a decision maker, at the instant of choice, is committed to having views about the character of her more fully considered opinions and preferences. When an agent, who has settled on some small-world decision problem, decides that a certain small-world act A is better than any alternative she becomes tacitly committed to thinking that her rationale for doing A would hold up were she to deliberate on the matter until further reflection became moot. If it is essential to her rationale that she is at least as confident in one small-world state as another, or that she finds one small-world outcome strictly preferable to another, then she is also committed to the view that she would retain these attitudes even if she continued refining her views about her predicament until she understood her grand-world decision in all its detail. (This assumes, of course, that she takes her deliberative processes to be in good working order, so that they provide a reliable method for getting at the truth about what she ought to do.) The moral is that a rational small-world decision maker should always make choices in a state in which she feels confident that her evaluations of the relative merits of acts would not be overturned if she were to reflect on her predicament more fully.

As a consequence, we can think of a rational agent's attitudes toward the states, outcomes, and acts in a small-world decision problem as her *best estimates* of the attitudes that she would hold regarding those states, outcomes, and acts in the grand-world context. When a

---

[31] Harman (1973).

decision maker forms an opinion about the likelihood of some small-world state (grand-world event) E, or about its likelihood relative to another state, she implicitly estimates the strength of her grand-world belief in E and tries to conform her current belief to it. Likewise, when she adopts a view about the desirability of a small-world outcome O, or its value relative to another such outcome, she again makes an implicit estimate of the degree to which she would find O desirable in the grand-world context and tries to conform her current desire to it. When she judges that some small-world act A is optimal, she commits herself to thinking that she would judge A best among the small-world alternatives if she considered the matter from the grand-world perspective.

There is, of course, no guarantee that her views on these matters will be correct. The agent might well judge that A is optimal among her small-world options, be firmly convinced that she would retain this opinion even if she reflected on the matter more fully, and yet it still might be true that she would come to see her choice of A as misguided if she actually were to deliberate further. It is not a mark of irrationality to have one's less than fully considered judgments conflict with one's fully considered judgments. The most we can reasonably ask of an agent is that she reflect on the decision problem she faces until she has good reason to think that further reflection would not change her views (at least not change them enough to affect her decision). At that point her job is to stop deliberating and solve the problem.

Now, since the beliefs and desires that come into play in small-world decision making carry commitments to views about the ultimate results of further deliberation it should be clear that the theory of rational choice has a substantial burden to carry in its handling of small-world decision making. It will not do simply to say that an agent's "true" views are the ones she would hold in the grand-world context, and then to go on to provide a theory that pertains only to this highly refined scenario. What one needs in addition is (a) some explanation of what it takes for a small-world decision maker's estimates of her grand-world attitudes to be correct, and (b) an account of rationality that applies to both grand- and small-world decision making and that guarantees that any small-world decision maker who correctly estimates her grand-world attitudes, and who adheres to the laws of rationality, will make a small-world choice that is rational when viewed from either the grand-world or small-world perspective. This is the *problem of small worlds*. It turns out to be rather difficult to solve. One theory that has a hard time solving it is that of Savage, which is next on our agenda.

3

Savage's Theory

In this chapter we investigate the version of expected utility theory developed in Savage's *The Foundations of Statistics*. In this immensely influential work, Savage sought to justify the principle of expected utility maximization by deriving it from plausible theses about the nature of rational preference. He did this by laying down a small system of axiomatic constraints on preference rankings, and then proving a general mathematical result, known as a *representation theorem*, which showed that any decision maker whose preferences satisfy the constraints automatically ranks acts by increasing expected utility. Savage took his axioms to express norms of rationality that could be justified without invoking any prior commitment to expected utility maximization. He thus saw his representation theorem as providing a rationale for such a commitment.

Savage was not the first person to prove a representation theorem of this type. The earliest such result may be found in (Ramsey 1931). Representation theorems had also appeared in (de Finetti 1964) and in (von Neumann and Morgenstern 1953) before *The Foundations of Statistics* was written. Savage's theorem, however, is better worked out, is more generally applicable, and has been far more influential than any of the rest.[1] Even though the theory has some limitations, it remains the best place to start when thinking about the foundations of decision theory.

The version of Savage's theory to be outlined here will not be identical in every respect to the one in *The Foundations of Statistics*, but it will still be recognizably Savage's. Most divergences between the two presentations are stylistic (e.g., I present the axioms in a slightly different order). All substantive differences can be traced to the fact that I am construing states, outcomes, and acts as propositions in the manner described in the last chapter, whereas Savage took states and outcomes as primitives and defined acts as functions from the former to the latter.

---

[1] This is partly a matter of his "standing on the shoulders of giants" like Pascal, Bernoulli, Ramsey, de Finetti, von Neumann and Morgenstern, and partly a matter of being a giant himself.

Savage sought to clarify and justify the expected utility hypothesis by first imposing a set of axiomatically specified constraints on rational preference, and then proving that these could be satisfied *only* by an expected utility maximizer. The theory was meant to apply to an agent facing a decision $D = (\Omega, O, S, A)$ where the agent's preferences over acts in A are given by a ranking ($>$, $\geq$). The set A is assumed to be highly structured. It is composed of bundles of Savage conditionals of the form $A = \&_s (S \Rightarrow O[A, S])$ and is required to contain all constant acts and to be closed under mixing as explained in the previous chapter. What Savage wanted to show was that, by obeying certain axiomatic constraints on preference an agent facing such a problem would automatically wind up maximizing expected utility in the sense that

- Her *degrees of confidence* regarding states in S and events in $\Omega(S)$ could be measured by a finitely additive subjective probability function P;
- The strengths of her basic desires for outcomes in O could be measured by a utility function u;
- The pair (P, u) defines an expected utility function that *represents* her preference over acts in the sense that, for all $A, A^* \in A$,

$$A > A^* \text{ only if } U(A) > U(A^*)$$
$$A \geq A^* \text{ only if } U(A) \geq U(A^*)$$

where U, the expected utility operator for u, is governed by

Savage's Equation: $U(A) = \Sigma_s P(S)u(O[A,S]).$[2]

Savage actually proved more than this. His axioms were strong enough to guarantee that "only if" can be replaced by "if and only if" in these expressions, that the probability P will be unique, and that the utility u will be unique up to the arbitrary choice of a unit and a zero point. What he showed, in other words, was that a decision maker whose preferences conform to his axioms will evaluate prospects *as if* she is maximizing expected utility relative to a unique P and an almost unique u.

---

[2] Notice how the "division of labor" among acts, states, and outcomes is reflected in this equation. Subjective probability is defined only for states in S and events in $\Omega(S)$, which, recall, are supposed to be the objects of the decision maker's uncertainty. Similarly, the utility function u is defined only over outcomes, the objects of her basic, noninstrumental desires. These together define the expected utilities of acts in A, which are the objects of instrumental desire.

Before we describe Savage's formalism in detail, it will be useful to spend a moment trying to understand why it is so important for proponents of expected utility maximization to prove representation theorems. One way to think about the issue is by focusing on the "as if" in the last sentence of the previous paragraph. No sensible person should ever propose expected utility maximization as a *decision procedure*, nor should he suggest that rational agents must have the maximization of utility as their *goal*. (Opponents of utility theory often misrepresent it as requiring these things, but it does not.) The expected utility hypothesis is a theory of "right-making characteristics" rather than a guide to rational deliberation. It in no way requires an agent *consciously* to assign probabilities to states of the world or utilities to outcomes, or to actually calculate anything. The decision maker does not need to have a concept of utility at all, and she certainly does not have to see herself as an expected utility maximizer. The demand is merely that her desires and beliefs, however arrived at, should be *compatible* with the expected utility hypothesis in the sense that it should be possible for a third party who knows her preference ranking to represent it in the way described.

This leaves an important open question. Even if we recognize that rational agents need not employ expected utility maximization as an explicit decision procedure, we do not, as yet, know what it *does* require of them. It is of little use to be told that one's preference ranking must be capable of being represented by an expected utility function unless this very "global" advice is accompanied by specific "local" recommendations that apply at the level of individual preferences. This is where representation theorems come in. One of the main purposes of these results is to show precisely what expected utility theory demands of agents by describing a system of "local" constraints on preferences, the satisfaction of which makes for global representability. These local constraints are what should guide a person in her evaluation of prospects. She ought to use them, as Savage puts is, "to police [her] decisions for consistency and, where possible, to make complicated decisions depend on simpler ones."[3] Thus, the direct obligation that the theory imposes is not that of maximizing expected utility, but that of conforming one's preferences to the axioms. A representation theorem will show that a person who meets her obligations in this regard will succeed in having beliefs and desires that are consistent with the global requirement of expected utility maximization.

[3] Savage (1954/1972, p. 20).

An analogy with deductive consistency might help illuminate these points. Most epistemologists agree that an epistemically rational agent must hold a corpus[4] of full beliefs that is globally consistent in the sense that it should be possible for all the believed propositions to be true together. If we think of a state of full belief as one of accepting some sentence as true (admittedly an imperfect model), then one might plausibly cash out this requirement by saying there should be some Tarskian interpretation of the agent's language that makes all of the sentences she believes true. Now, no serious epistemologist will propose that people use *this* principle to run their doxastic lives. It would be a very bad idea to have us policing our beliefs by trying to construct Tarskian models that make them all true. Instead, rational agents should seek to conform their beliefs to rules that impose *local* consistency constraints like the following:

- **The Law of Noncontradiction.** Do not believe $X \ \& \ \neg X$ for any sentence X.
- *Modus ponens.* Do not believe X, $X \supset Y$, and $\neg Y$ for any sentences X and Y.

These rules of local consistency state necessary conditions for global consistency. They are the norms that a rational full believer will use as guides in determining what propositions to accept. In thinking about what to believe one is generally led not by the overarching prescription to hold a corpus of belief that has a Tarskian model, but by myriad more modest and manageable prescriptions like the two listed.

This is not merely a practical point; one does not use the local rules solely because they are easier to apply. Rather, they are part of what gives the global consistency requirement its bite. The soundness and completeness theorems show that a set of sentences has a Tarskian model if and only if it does not violate any of the local consistency constraints embodied in the axioms of the standard predicate calculus. Thus, one cannot commit the global epistemic sin of having a corpus of beliefs that lacks a Tarskian model without thereby violating some local rule like *modus ponens*. Likewise, a violation of any of the local constraints is sure to result in a violation of the global one. In this way, the completeness and soundness theorems tell us what the requirement of global consistency demands at the "local level" by showing us that satisfying a certain system of local constraints is necessary and sufficient for achieving the optimal global state. This opens up certain

---

[4] I am taking the term from Levi (1980).

justificatory possibilities. A person can, if so inclined, justify the global consistency requirement to believe a set of sentences that has a Tarskian model by appealing to the independent plausibility of the local constraints. Alternatively, she can justify local constraints like *modus ponens* by appealing to the independent plausibility of the global requirement. Or, if she feels that both the global and local requirements have a great deal of independent plausibility (as seems reasonable in this case), she can invoke the completeness and soundness theorems as a part of a kind of symbiotic justification in which each requirement is supported by its association with the other.

A perfect representation theorem would do for rational desire what the soundness and completeness theorems do for rational full belief. That is, it would lay down a small set of axioms that deem certain local patterns of preference rational, and others irrational, and would show that these axioms are individually necessary and jointly sufficient for the existence of the desired expected utility representation. This would help us understand what the global mandate to maximize expected utility demands at the level of individual preferences. It would also make it possible for proponents of expected utility maximization to rest their case on the plausibility of the local axioms rather than the expected utility principle itself (provided that the axioms can be justified independently), or to offer a symbiotic justification of the local and global requirements by appealing to the independent plausibility of each.

Unfortunately, no currently available theorem, including Savage's, quite succeeds in doing this because all of them secure representability by including axioms that are not strictly required for the existence of the desired expected utility representation. These nonnecessary axioms are existential statements that concern the size and complexity of preference rankings. I will follow Patrick Suppes in calling them *structure axioms*.[5] The other, necessary axioms are *axioms of pure rationality*. These are universally quantified statements that do not force preference rankings to be defined over large or complex sets of propositions, but that cannot be violated if the desired representation is to exist. In presenting Savage's theorem I will be careful to distinguish the structure axioms from the rationality axioms by placing a "*" in front of the former when they are introduced. After we see how his system works we will ask whether it is possible to avoid the use of structure axioms altogether.

[5] Suppes (1984).

Before we begin presenting Savage's axioms we should remind our-selves of two points having to do with his conception of decision problems. Recall first that Savage assumed that the acts in **A** would be the *only* things the agent finds instrumentally desirable. As a conse-quence, his entire theory is framed solely in terms of the agent's preferences over acts. Second, we should remember that **A** has a great deal of structure. To call attention to it, it will be useful to include an axiom that does not figure in Savage's official theory, but that is tantamount to his explicit, but unaxiomatized, identification of acts with abstract functions from states to outcomes.

*SAV_0 (Act Richness): For any outcome $O \in$ **O**, **A** contains the "constant" act $A_O = \&_S (S \Rightarrow O)$ that has O as its outcome in every state of the world. Moreover, if A and B are in **A** and if E is any event, then the "mixed" act

$$A_E \& B_{\neg E} = \&_S [(S \& E \Rightarrow O[A,S]) \& (S \& \neg E \Rightarrow O[B,S])]$$

is also an element of **A**.

SAV_0 says that the set of acts contains all constant acts and is closed under "mixing." The reader will want to compare this axiom with Premise 1 in the argument given for Pascal's Thesis in Section 1.4.

For future reference we should note that it is a consequence of SAV_0 that **A** contains all propositions that can be constructed out of constant acts by a finite number of applications of mixing. Let **G** be the set of all these propositions, so that **G** = $\{A^1_{E1} \& A^2_{E2} \& \ldots \& A^n_{En}$: each A^j is a constant action and $\{E_1, E_2, \ldots, E_n\}$ a finite partition of events in $\Omega(S)\}$. **G** is the set of all *wagers* in **A**. (Savage uses *gambles*.) In Chapter 1 a wager was defined as an arrangement in which a bettor will be paid a certain sum of money depending upon which one of a set of mutually exclusive jointly exhaustive event propositions is true. The situation here is the same except that the "payoffs" are not necessarily monetary ones. Most of Savage's axioms apply to an agent's preferences for wagers.

The next axiom in our presentation of Savage's theory (his postu-late P5) says that it must be possible for the truth of propositions in **A** to make some difference to the decision maker's happiness.

*SAV_1 (Nontriviality): There are outcomes $O, O^* \in$ **O** such that the agent strictly prefers the truth of the constant act A_O to that of the constant act A_{O^*}.

This should be uncontroversial. For SAV_1 to fail it would have to be a matter of indifference to the agent which outcome proposition is true. She would have achieved the state that the Buddhists call nirvana, in which every prospect she considers seems equally desirable from her point of view. Such a person has no decisions to make, and it in no way diminishes the theory of rational choice to admit that it does not apply to her.

The second axiom requires the decision maker's preferences to partially order the propositions in the base algebra Ω. We actually do not need such a strong requirement to state Savage's theory; a partial order over **A** would suffice. It does no harm to impose the more general requirement, and doing so will make it easier for us to extend the theory in future chapters.

SAV_2 (Partial Ordering): The decision maker's preferences *partially order* the propositions in Ω; that is for any $X, Y, Z \in \Omega$ one has

> *Reflexivity of Weak Preference*: $X \geq X$
> *Irreflexivity of Strict Preference*: Not $X > X$
> *Consistency*: $X > Y$ only if $X \geq Y$
> *Transitivity:* If $X \geq Y$ and $Y \geq Z$ then $X \geq Z$
> If $X > Y$ and $Y \geq Z$ then $X > Z$
> If $X \geq Y$ and $Y > Z$ then $X > Z$

The only aspect of this axiom that is even moderately controversial is the transitivity clause. In the next section, we will consider some alleged counterexamples to transitivity, all of which will be seen to be based on the same misconception. Apparently rational intransitive preferences are nearly always a sign that the agent lacks a definite preference. Most objections that are offered against transitivity should really be aimed at the next axiom.

As noted, Savage assumes that ($>$, $\geq$) constitutes a *complete* order of the set of acts **A** (and thus **O** as well). He does so by introducing a single axiom (P1 in his system) that combines SAV_2 and

*SAV_3 (Completeness): The decision maker has a definite preference with regard to every pair of propositions in **A**, so that either $A > B$ or $B \geq A$ holds for every $A, B \in$ **A**.

I have chosen to separate *SAV_3 from SAV_2 to highlight the fact that the two axioms have a different character. As the "*" indicates, *SAV_3 is a structure axiom; it imposes a constraint on the agent's preferences that is not strictly required for the existence of an expected utility representation. It is a good thing that it is not required since, given how large and complex the set **A** turns out to be, it would be unreasonable to require a decision maker to have a definite preference regarding each pair of its elements. We will see later how to get by without this requirement.

The next two axioms express different aspects of the famous *sure-thing principle*. This principle, which serves as the cornerstone of Savage's entire theory, is what ultimately ensures that rational preference rankings can be represented by expected utility functions that are "linear in the probabilities." Savage introduces it using the following example:

A businessman contemplates buying a certain piece of property. He considers the outcome of the next presidential election relevant to the attractiveness of his purchase. So, to clarify the matter for himself, he asks whether he would buy if he knew that the Republican candidate were going to win, and decides that he would do so. Similarly, he considers whether he would buy if he knew that the Democratic candidate were going to win, and again finds that he would do so. Seeing that he would buy in either event [and assuming that no third party candidate can win], he decides that he should buy, even though he does not know which event obtains.[6]

Savage then goes on to suggest a general rationale for the businessman's choice:

If [he] would not prefer A to B, either knowing that the event E obtained or knowing that the event $\neg E$ obtained, then he [ought] not prefer A to B. Moreover (provided that he does not regard E as virtually impossible) if he would definitely prefer A to B knowing that E obtained, and, if he would not prefer B to A, knowing that E did not obtain, then he will definitely prefer A to B.[7]

There are really two different principles here, one in the first sentence and one in the second.

With the help of conditional acts we can state the first part of the sure-thing principle as follows: For any acts A and B and any event E, if $A_E \geq B_E$ and $A_{\neg E} \geq B_{\neg E}$ then $A \geq B$. In the presence of the other axioms this turns out to be equivalent to

SAV₄ (Independence): Suppose that A and A^* produce the same outcomes in the event that E is false, so that $A_{\neg E} = A^*_{\neg E}$. Then, for any act $B \in$ **A**, one must have

$$A > A^* \text{ if and only if } A_E \ \& \ B_{\neg E} > A^*_E \ \& \ B_{\neg E}$$
$$A \geq A^* \text{ if and only if } A_E \ \& \ B_{\neg E} \geq A^*_E \ \& \ B_{\neg E}{}^8$$

[6] Savage (1954/1972, p. 21).
[7] Savage (1954/1972, pp. 21–22), with minor changes in notation.

85

In other words, a rational agent's preference between A and A^* should not depend on what happens in circumstances where the two yield identical outcomes. There is a clear intuitive motivation for this principle. Given that A and A^* produce equally good results when E is false it is plausible to think that any reason for preferring one to the other would have to be based on an assessment of their relative merits *when E is true.* If this is right, then any reason that militates in favor of preferring A to A^* will also militate in favor of preferring $(A_E$ & $B_{\neg E})$ to $(A^*_E$ & $B_{\neg E})$. After all, these latter two acts differ from A and A^*, respectively, only in that they have replaced the shared outcome $O[A, S] = O[A^*, S]$ by $O[B, S]$ in each state S that is inconsistent with E. One way to state the point is to say that a rational agent should make judgments about the relative desirabilities of acts by treating their common outcomes as "dummy variables" whose effects "cancel out" in her deliberations.

To state the second part of the sure-thing principle we must introduce the notion of a *null* event. Intuitively, an event E is null when the decision maker is certain that it will not come about. Here is an example of an event that is null for me at the moment:

E: In the next five minutes the earth's atmosphere will spontaneously migrate to the southern hemisphere and remain there for all eternity unconstrained by any barrier.

The physicists say that events like E have no chance whatsoever of coming about, and I believe them. Thus, while I regard E's truth as logically possible, I am sure it is false. This will be manifested in my preferences by my complete disregard for outcomes that can only come about when E is true. (I am not, for example, writing this on a plane to Rio.) In Savage's theory the irrelevance of null events is captured by the following definition and axiom:

An event E is *null* if and only if there are $A, B \in \mathbf{A}$ such that the decision maker is indifferent between A_E & $A_{\neg E}$ and B_E & $A_{\neg E}$ even though she strictly prefers $O[A, S]$ to $O[B, S]$ for every state S that is consistent with E.

In other words, E is null when the fact that the agent is sure to do better with A_E & $A_{\neg E}$ than with B_E & $A_{\neg E}$ in the event that E is true

---

[8] One of these two conditions is redundant in the presence of SAV$_3$. I include them both because we shall be asking what happens when the nonnecessary condition SAV$_3$ fails.

makes no difference at all to her preferences. The next axiom requires the agent to be consistent in her treatment of null events.

SAV$_5$ (Nullity): If E is null, then the decision maker will be indifferent between A_E & $A_{\neg E}$ and B_E & $A_{\neg E}$ for any $A, B \in$ **A**.

In the presence of the other axioms **SAV$_5$** ensures that null events will have probability 0 in any probabilistic representation of the agent's beliefs.

SAV$_4$ and SAV$_5$ impose a kind of "noncontextuality" constraint on the decision maker's beliefs and values in that they do not allow her estimates of the likelihoods of states in **S** or her views about the desirabilities of the outcomes in **O** to depend on what member of **A** happens to be true. To see why this is necessary for the existence of Savage's expected utility representation, notice that we can write

$$U(A_E \ \& \ A_{\neg E}) = \Sigma_{S \epsilon E} P(S) u(O[A,S]) + \Sigma_{S \epsilon \neg E} P(S) u(O[A,S])$$

$$U(B_E \ \& \ A_{\neg E}) = \Sigma_{S \epsilon E} P(S) u(O[B,S]) + \Sigma_{S \epsilon \neg E} P(S) u(O[A,S])$$

Since these two expressions are identical to the right of the "+" sign, proponents of expected utility maximization are committed to thinking that neither the probabilities of states that are inconsistent with E nor the utilities of outcomes that these states bring about should matter to a rational agent's evaluation of A and B_E & $A_{\neg E}$. This can only happen if the agent's beliefs and desires about these states and outcomes do not vary depending upon whether A or B_E & $A_{\neg E}$ is true.

We will discuss the issue of probabilistic independence in the next two chapters. The best way to think about the independence of desires is by reflecting on the way in which conditional acts differ from many other propositions. As G. E. Moore once observed, most goods have a kind of "organic unity" that makes it impossible to assign them a value independent of the context in which they appear. One may, say, prefer wine to beer if one has cheese and crackers to eat, and yet prefer beer to wine if there are pretzels to be had. In terms of propositions, Moore's point is that there is nothing irrational *per se* about simultaneously preferring the truth X & Z to that of Y & Z, and also dispreferring the truth of X & Z^* to that of Y & Z^* because the agent's views about the relative merits of X and Y may depend on the context (Z or Z^*) in which she evaluates them. While expected utility theorists do not dispute Moore's general point, they do maintain that the principle of organic unity fails in the special case where X and Y

87

are acts conditional on E and Z is an act conditional on $\neg E$. SAV_4 and SAV_5 tell us that the amount of satisfaction an agent derives from the truth of A_E ought *not* to be a function of context at least when A_E is conjoined with propositions of the form $B_{\neg E}$. In other words, her desire for A_E's truth should never be altered by considerations pertaining to how things would go if E were false (since such considerations cannot effect the likelihood or desirability of the outcomes that A_E might produce). This is not how Savage would put things, since he did not assume that his decision maker would have preferences for conditional acts, but it is an illuminating way to clarify the content of his axiom.

Savage used SAV_4 and SAV_5 to define the agent's *conditional preference* ranking. This is a family of binary relations ($>_E$, $\geq_E$), indexed by nonnull events in Ω, that characterizes the decision maker's views about the desirabilities of various acts in **A** under the supposition that specific events obtain. Here is the official definition:

> An agent whose unconditional preferences are given by ($>$, $\geq$) strictly/weakly prefers A to B conditional on a nonnull event E, written $A >_E B$ and $A \geq_E B$, if and only if she strictly/weakly prefers A_E & $C_{\neg E}$ to B_E & $C_{\neg E}$ for some (hence, by Independence, for every) $C \in$ **A**.

SAV_5 ensures that ($>_E$, $\geq_E$) is defined for nonnull E, but leaves it undefined when E is null.

Savage's next axiom requires the decision maker's preferences to make probabilistic sense. Let O^A and O^B be outcomes in **O** such that the agent strictly prefers O^A to O^B, and let A and B be the "constant" acts associated with O^A and O^B, respectively. *The wager on E with the difference between O^A and O^B at stake* is the act A_E & $B_{\neg E}$ whose outcomes are exactly as desirable as O^A when E holds and exactly as desirable as O^B when E does not hold. The next axiom says that a rational decision maker's comparative preferences among wagers of this sort should not depend on the size of the stake (as long as it is the same in both cases).

SAV_6 (Stochastic Dominance): Let O^A, O^B, O^C, and O^D be outcomes in **O** such that the decision maker strictly prefers O^A to O^B and O^C to O^D. Suppose also A, B, C, and D are the "constant" acts associated with O^A, O^B, O^C, and O^D. Then, for any events E and F, one should have

$$A_E \ \& \ B_{\neg E} > A_F \ \& \ B_{\neg F} \text{ if and only if } C_E \ \& \ D_{\neg E} > C_F \ \& \ D_{\neg F}$$
$$A_E \ \& \ B_{\neg E} \geq A_F \ \& \ B_{\neg F} \text{ if and only if } C_E \ \& \ D_{\neg E} \geq C_F \ \& \ D_{\neg F}$$

This is the analogue of the Stochastic Dominance lemma of Chapter 1. It says that an agent who would prefer to stake the difference between O^A and O^B on E rather than F should also prefer to stake the difference between O^C and O^D on E rather than F.

The intuitive appeal of this principle should be obvious. It simply says that an agent should favor prospects that, by her own lights, offer her a greater chance of obtaining more desirable outcomes. Someone who cares only about outcomes, as we are assuming our decision maker does, would have no reason to favor A_E & $B_{\neg E}$ over A_F & $B_{\neg F}$ unless she thinks that the former prospect is more likely to bring about O^A. And, if she does think this, she also has a reason for favoring C_E & $D_{\neg E}$ over C_F & $D_{\neg F}$ since here too the first wager is more likely than the second to yield the desirable result. Following Savage, we can say that the agent *regards E as more likely than (as likely as) F*, written $E \ldots F$ ($E .\geq. F$), just in case she strictly (weakly) prefers A_E & $B_{\neg E}$ to A_F & $B_{\neg F}$. The pair $(.>. , .\geq.)$ constitutes the agent's *comparative likelihood ranking*. The principal purpose of SAV_6 is to ensure that this ranking is well-defined. More formally, it guarantees

Coherence. For *any* constant acts such that $A > B$ one has

$E .>. F$ if and only if A_E & $B_{\neg E} > A_F$ & $B_{\neg F}$
$E .\geq. F$ if and only if A_E & $B_{\neg E} \geq A_F$ & $B_{\neg F}$

This (partially) characterizes the relationship that must hold between a preference ranking over **A** and a comparative likelihood ranking over $\Omega(\mathbf{S})$ for the two to be candidates for joint representation by a (P, u) pair obeying Savage's Equation.

Let's pause a moment to clear up a potential source of confusion about the status of $(.>. , .\geq.)$. In keeping with his behaviorism, Savage treated Coherence as a kind of *operational definition* of the notion of comparative belief. To be more confident in E than F, he thought, *just is* to have the given pattern of preferences (which are supposed to reveal themselves in actions). This is the wrong way to view things. First, behaviorism is untenable for reasons we have already discussed. More to the point though, even among decision theorists who would reject behaviorism, the idea that Coherence is a *definition* of . . . and .$\geq$. can foster a kind of pragmatism that sees belief as a second-class propositional attitude that can only be understood in terms of its relationship to desire. Teddy Seidenfeld and M. Schervish have given a clear statement of the view I have in mind (without endorsing it):

One of the most important results of the theories of Savage and de Finetti is the thesis that, normatively, preference circumscribes belief. Specifically, these authors argue that the theory of subjective probability is reducible to the theory of reasonable preference, i.e., coherent belief is the consequence of rational desire.[9]

The view being expressed here can be succinctly stated as

Pragmatism. The laws of rational belief are underwritten by the laws of rational desire. To call any belief rational or irrational is always to say something about the rationality of the believer's desires.

To put an even finer point on it, the claim is that we can learn everything we need to know about epistemology by doing decision theory.

It is important to understand that the formalism of decision theory does not *force* us into this view. Rather than regarding Coherence as a definition of $(.>., .\geq.)$ we can see it as a description of a relationship between two coequal propositional attitudes that are, to some extent, governed by different laws. While I will not argue for this in detail here, I think this is the correct view of the matter. Elsewhere[10] I have suggested that the laws of rational belief are ultimately grounded not in facts about the relationship between belief and desire, but in considerations that have to do with the pursuit of truth. No matter what our practical concerns might be, I maintain, we all have a (defeasible) epistemic obligation to try our best to believe truths as strongly as possible and falsehoods as weakly as possible. Thus, we should look to epistemology rather than decision theory to find the laws of rational belief. Just to make the point clear, I am not denying that we can learn important and interesting things about the nature of rational belief by considering its relationship to rational desire and action. What I am denying is the radical pragmatist's claim that this is the *only*, or even the most fruitful, way to approach such issues.

We now come to the crucial step of Savage's theorem. To secure the existence of an expected utility representation for the preference ranking $(>, \geq)$ Savage needed to find a probability function P, defined on $\Omega(\mathbf{S})$, that would represent the comparative likelihood ranking $(.>., .\geq.)$ in the sense that

$$E .>. E^* \text{ only if } P(E) > P(E^*)$$
$$E .\geq. E^* \text{ only if } P(E) \geq P(E^*)$$
$$E \text{ is null only if } P(E) = 0$$

[9] Seidenfeld and Schervish (1983, p. 398).
[10] Joyce (1998).

He began by showing that the axioms presented thus far ensure that $(.>., .\geq.)$ satisfies Bruno de Finetti's *laws of comparative probability*.[11]

CP$_1$ (*Normalization*): Tautologies are more likely than contradictions, so that $E \vee \neg E .>. E \& \neg E$ for any event E.

CP$_2$ (*Boundedness*): A tautology is at least as likely as any event, and any event is as likely as a contradiction, that is, $E \vee \neg E .\geq. E .\geq. E \& \neg E$ for every event E.

CP$_3$ (*Ranking*): $(.>., .\geq.)$ is a partial ordering of $\Omega(\mathbf{S})$.

*CP$_4$ (*Completeness*): Any two events are comparable with respect to likelihood, so that $E .>. E^*$ or $E^* .\geq. E$ for any $E, E^* \in \Omega(\mathbf{S})$.

CP$_5$ (*Quasi-additivity*): If F is incompatible with both E and E^*, then

$$E .>. E^* \text{ if and only if } E \vee F .>. E^* \vee F$$
$$E .\geq. E^* \text{ if and only if } E \vee F .\geq. E^* \vee F$$

The key axiom is the last one. It is a qualitative analogue of the additivity law of probability. Among other things, it entails that the agent will always judge an event to be at most as likely as anything it entails, and that the set $\mathbf{N}$ of all events she judges to be as likely as a contradiction is an *ideal* (that is, a set that contains the contradictory proposition, that is closed under countable disjunction, and that contains $E \& F$ for any $F \in \Omega(\mathbf{S})$ whenever it contains E).

The result Savage proved was this:[12]

Lemma 3.1. *Suppose that $(>, \geq)$ and $(.>., .\geq.)$ jointly satisfy Coherence. A necessary condition for the preference ranking $(>, \geq)$ to obey SAV_0–SAV_6 is that the comparative likelihood ranking $(.>., .\geq.)$ must both obey CP_1–CP_5 and have a null ideal $\mathbf{N}$ that coincides with the set of null events determined by $(>, \geq)$.*

The proof will be left to the reader. (The key step is to use the SAV$_4$

---

[11] De Finetti (1964). The version of these laws given here is slightly different from what one finds in de Finetti's work, but the differences are superficial matters of presentation.

[12] Pragmatists will want to portray Lemma 3.1 as *justifying* the claim that a rational agent's comparative likelihood ranking should satisfy de Finetti's axioms. My nonpragmatist inclination is to run the inference in the reverse, and to see it as a point in favor of Savage's axioms that they entail such obvious truths about comparative beliefs.

to derive CP_5.[13]) The reader is also invited to verify that CP_1, CP_2, CP_3 and CP_5 are each necessary for the existence of a probability representation for $(.>., .\geq.)$, and that $*CP_4$ is necessary if the representation is going to have the biconditional character Savage wants.

If the set of events is infinite, as it will often be, the following additional conditions must be imposed if probabilistic representability is to be ensured:

CP_6 (*Archimedean Axiom*): If C is a partition of nonnull events in $\Omega(S)$ such that $E .=. E^*$ for every $E, E^* \in C$, then C is finite.

CP_7 (*Continuity*): If $\{E_1, E_2, E_3, \ldots\}$ is a countable set of mutually incompatible events in $\Omega(S)$ and if $E^* .\geq. (E_1 \vee E_2 \vee \ldots \vee E_n)$ for all n, then $E^* .\geq. \vee_j E_j$.

CP_6 rules out infinitesimally small probabilities. CP_7 is a qualitative expression of the fact that, when E_1, $E_2, \ldots$ are contraries, the sequence $P(E_1 \vee \ldots \vee E_n)$ converges to $P(\vee_i E_i)$ from below as n approaches infinity. This is necessary for the existence of a countably additive representation.

De Finetti initially thought that his axioms would ensure that $(.>., .\geq.)$ has a finitely additive representation. This turned out to be false. As C. Kraft, J. Pratt, and A. Seidenberg showed in 1959, there are comparative likelihood rankings that satisfy CP_1–CP_5 and cannot be represented by any probability function.[14] (Adding CP_6 and CP_7 is no help because the counterexample of Kraft et al. is based on a *finite* algebra of events.) To secure representability one must either impose stronger qualitative conditions on $(.>., .\geq.)$ or require it to obey certain, nonnecessary structural conditions.

Savage (who did not know about the results of Kraft et al.) adopted the latter approach; he proposed that $(.>., .\geq.)$ should obey the following condition:

$*CP_8$ (*Richness of $\Omega(S) \sim N$*): If $E .>. E^*$, then there exists a finite partition $\{C_1, C_2, \ldots, C_n\}$ of nonnull events such that $E .>. (C_j \vee E^*)$ for every C_j.

[13] Here is how to show that $E .>. E^*$ iff $(E \vee F) .>. (E^* \vee F)$:

$(A_E \& B_{\neg E}) > (A_{E^*} \& B_{\neg E^*})$	(definition of $E .>. E^*$)
$(A_E \& B_F \& B_{\neg E \& \neg F}) > (A_{E^*} \& B_F \& B_{\neg E^* \& \neg F})$	(since F is a contrary of E and E^*)
$(A_{E \vee F} \& B_{\neg E \& \neg F}) > (A_{E^* \vee F} \& B_{\neg E^* \& \neg F})$	(by Independence)

Since A and B are constant acts with $A > B$, this is equivalent to $(E \vee F) .>. (E^* \vee F)$.

[14] Kraft et al. (1959).

This implies that the algebra of events is infinite and that the ranking $(.>., .\geq.)$ is *atomless* in the sense that each nonnull event can be partitioned into a pair of disjoint nonnull events. Once $*CP_8$ was added to CP_1–CP_5 Savage had the resources to establish the existence of a finitely additive probability representation for $(.>., .\geq.)$ and he was able to show that this representation would be unique.

Lemma 3.2.[15] *If the comparative likelihood ranking $(.>., .\geq.)$ satisfies CP_1–CP_5 and $*CP_8$, then there exists a unique finitely additive probability P defined on $\Omega(\mathbf{S})$ such that*

$$E .>. E * \text{ if and only if } P(E) > P(E *)$$
$$E .\geq. E * \text{ if and only if } P(E) \geq P(E *)$$

One should notice that the P that Savage constructs need only be *finitely* additive. C. Villegas was able to prove that adding the continuity principle CP_7 forces P to be countably additive as well.[16] For future reference, note that Villegas also showed that, in the presence of CP_7, $*CP_8$ can be replaced by the weaker claim that $(.>., .\geq.)$ is atomless. This fact will be important in the sequel.

To make sure that $(.>., .\geq.)$ would obey $*CP_8$ Savage imposed a further structure axiom on the decision maker's preferences:

$*SAV_7$ (*Event Richness*): If $A > B$ and C is any constant act, then there is a finite partition of events such that both $(A_E \, \& \, C_{\neg E}) > B$ and $A > (B_E \, \& \, C_{\neg E})$ hold for any event E in the partition.

This is a powerful axiom. For any outcome $O \in \mathbf{O}$ and any $A, B \in \mathbf{A}$, it requires that there be a partition of nonnull events whose probabilities are so small that it does not materially affect the agent's preference for A over B when either option is made exactly as desirable as O on any event in the partition. For this to hold two things must occur. First, each element of $\Omega(\mathbf{S}) \sim \mathbf{N}$ must contain subevents of arbitrarily low finite probability. Second, there can be no outcomes in $\mathbf{O}$ that are so desirable or so undesirable that their effects cannot be "canceled out" by making their probabilities small enough. The first point, as Savage notes, is easy to imagine satisfied. If there is a coin, whose chance of coming up heads falls between 0 and 1, that can be flipped an arbitrary number of times, then for any event E there will be a subevent $E* = (E \, \& \, \text{The coin comes heads } k \text{ times in } n \text{ tosses.})$ that might have any finite rational-valued probability. The second

---

[15] The proof is rather messy. Interested readers may consult Savage (1954/1972, pp. 31–40) for details.

[16] Villegas (1964).

requirement is more substantive. It rules out the possibility of goods, like the reward of heaven in Pascal's wager, whose value is so great that no improbability can diminish them.

An example may help to clarify this last point. Suppose that your immediate death is a good approximation to a constant act from your point of view, and let A and B describe situations in which you stand to receive \$1 or \$0 (other things being equal). Imagine further that there are an infinite number of revolvers set out before you: one has a single chamber, one has two chambers, one has three, and so on for every natural number n. Each gun contains a single bullet. Let me offer you the chance to play Russian roulette with any gun you choose where the prize for playing and living is A, the prize for not playing is B, and the prize for playing and losing is death. Will you play? SAV_8 says that you should. More precisely, it says that there should be some number n such that you are willing to bet your life against a dollar in a game of Russian roulette with a revolver having one bullet in n (or more) chambers. So, if you prefer a dollar to nothing, then you cannot value your life so much that you are not willing to risk it in order to win a dollar at some positive odds. If you are inclined to shy away from this conclusion, as many people are, reflect on the fact that walking into a bank to withdraw money exposes you to a *much* greater risk of being shot to death than playing Russian roulette with a re-volver having as few as a billion chambers.[17]

The theory developed thus far allowed Savage to prove a represen-tation result that is almost as general as the one he sought. What Savage showed was that SAV_1–SAV_7 make it possible to find an expected utility representation of the agent's preferences over the

---

[17] People sometimes object to this by pointing out *they* would not be shooting *them-selves* in a bank robbery, and that they assign a high value to not committing suicide. Indeed, some even profess to believe that it is infinitely better to be done in by a bank robber than to die by one's own hand, so that no amount of money would be worth risking the latter. No one really thinks this, though. People do, after all, go into banks to withdraw money all the time, even though there is some (very tiny, but nonzero) probability that doing so will begin a sequence of events that will make killing themselves a *desirable* option. I do not mean to be offering a pure *ad hominem* here. The point is to remind us that we all have a highly ingrained tendency to overestimate the chances of events that have extremely good or extremely bad outcomes. This propensity is well documented in the psychological literature. Shafir and Tversky (1995) is an excellent survey. It explains, for example, why people play the lottery, and why they overpay for insurance. The intuition that there is no probability at which one would risk one's life for a dollar is an instance of this tendency to overestimate the probabilities of extreme events, a tendency we *all* recognize as irrational in a wide variety of cases. Thus, the point of the *ad hominem* attack is to help people see their preferences here as instances of a very general and well-documented kind of irrationality to which we are all susceptible in ordinary, non-life-or-death cases.

subset **G** of **A** that contains all wagers. I will not give the proof here, but a quick sketch of its basic structure may be helpful. After using Lemma 3.1 and Lemma 3.2 to establish the existence of a probability P that represents ($.>.$, $.\gtrsim.$), Savage arbitrarily chose two constant actions C and C^* with $C > C^*$ and set $u(C^*) = 0$ and $u(C) = 1$ to fix a unit and a zero relative to which utility can be measured. He then showed that the richness and the ordering axioms (SAV$_7$ and SAV$_2$) entail that exactly one of the following will be true of any *constant* act A not ranked with C or C^*:

- $C > A > C^*$ and there are nonnull E and $\neg E$ with $A \approx (C_E \,\&\, C^*_{\neg E})$.
- $A > C > C^*$ and there are nonnull E and $\neg E$ with $C \approx (A_E \,\&\, C^*_{\neg E})$.
- $C > C^* > A$ and there are nonnull E and $\neg E$ with $C^* \approx (C_E \,\&\, A_{\neg E})$.

The expected utility hypothesis forces one to set the value of $u(A)$ at $P(E)$, $1/P(E)$, and $- P(E)/P(\neg E)$ in these three cases, respectively. This defines a utility over all the *constant* acts in **A**. It is an instructive exercise to convince oneself that this assignment ordinally represents the decision maker's preference ranking over constant acts. Savage used the other axioms, notably Independence (SAV$_4$) and Stochastic Dominance (SAV$_6$), to show that the restriction of ($>$, $\geq$) to **G** obeys all the von Neumann-Morgenstern axioms relative to the probability P and the utility u. It then follows from the von Neumann-Morgenstern theorem (Theorem 1.3) that the function defined by

$$U\!\left(A^1_{E1} \,\&\, A^2_{E2} \,\&\dots\&\, A^n_{En}\right)$$

$$= P(E_1)u\!\left(A^1\right) + P(E_2)u\!\left(A^2\right) + \dots + P(E_n)u\!\left(A^n\right)$$

is an expected utility representation of the agent's preferences for elements of **G**.

To extend this representation to all of **A** Savage needed to generalize the sure-thing principle to cover acts other than finite wagers. I will not state the required axiom in quite the form Savage did, but its content is the same.

SAV$_8$ (Averaging): It is not the case that $A >_E O[A, S]$ for all states S consistent with E nor that $O[A, S] >_E A$ for all S consistent with E.[18]

This says that the agent must judge that A's desirability given E falls between that of its highest and lowest possible outcomes given E,

---

[18] I am speaking somewhat sloppily here. In Savage's system $A >_E O[A, S]$ does not make sense. It really should be read as saying that, conditional on E, the agent prefers A to the constant act whose outcome is always $O[A, S]$.

which is the only way it could be, given that she is assumed to be evaluating acts in terms of the outcomes they produce.

By adding this axiom Savage was able to extend his expected utility representation for (**>**, **≥**) from **G** to all of **A**. Here is the result:

Theorem 3.3 (Savage's Representation Theorem). *Suppose that an agent, whose desires are described by the preference ranking (**>**, **≥**), faces the decision problem **D** = (Ω, **O**, **S**, **A**). If (**>**, **≥**) satisfies all the Savage axioms, then*

(A) *there exists a unique finitely additive subjective probability function P defined over events in Ω(**S**) such that*
   - $P(E) = 0$ if and only if $(A_E \, \& \, B_{\neg E}) \approx B$
   - $P(E) > P(F)$ if and only if $A_E \, \& \, B_{\neg E} > A_F \, \& \, B_{\neg F}$
   - $P(E) \geq P(F)$ if and only if $A_E \, \& \, B_{\neg E} \geq A_F \, \& \, B_{\neg F}$
   for all constant acts A and B with A > B.

(B) *there exists a bounded utility function u defined on **O**, which is unique up to the choice of a zero and a unit, such that for all A, A* ∈ **A**, one has*
   - $A > A^*$ exactly if $\Sigma_S P(S)u(O[A, S]) > \Sigma_S P(S)u(O[A^*, S])$
   - $A \geq A^*$ exactly if $\Sigma_S P(S)u(O[A, S]) \geq \Sigma_S P(S)u(O[A^*, S])$.

One obvious formal limitation of this theorem is that it only provides for the existence of a *finitely* additive probability to represent (.>. , .≥.). Savage was not bothered by this. He wrote, "however convenient countable additivity may be, it, like any other assumption, ought not be listed among the postulates . . . unless we feel that its violation deserves to be called inconsistent or unreasonable."[19] While Savage seems not to have such feelings, it seem clear to me that an agent whose likelihood ranking lacks a countably additive representation is going to have beliefs and desires that are unreasonable. Among other things, her beliefs are going to be *non-conglomerable* in the sense that there will be occasions on which her probability for E conditional on each element of a countable partition $H_1, H_2, H_3, \ldots$ will fall below some value x, so that $P(E/H_j) = P(E \, \& \, H_j)/P(H_j) < x$ for all j, and yet E's probability will come in above x.[20]

This is not an insurmountable problem for Savage's theory. As Peter Fishburn has shown,[21] one can obtain a countably additive

[19] Savage (1972, p. 43).
[20] For an excellent discussion of non-conglomerability see Schervish et al. (1984).
[21] Fishburn (1970, p. 213).

96

representation by replacing SAV_8 with the slightly stronger, but still very reasonable, principle

SAV_8^+ (Dominance): If $O[A, S] > O[B, S]$ for all S consistent with the nonnull event E, then $A >_E B$. Likewise, if $O[A, S] \geq O[B, S]$ for all S consistent with the nonnull event E, then $A \geq_E B$.

This simply says that a rational agent will strictly prefer A to B given E if A's outcome is certain to be better than B's if E is true, and that she will weakly prefer A to B given E if it is impossible for A's outcome to be worse than B's if E is true. This requirement suffices to guarantee P's countable additivity. Its addition to SAV_1–SAV_7 completes the development of Savage's theory.

3.3 THE STATUS OF SAVAGE'S AXIOMS

One's view about the significance of Savage's theorem will, of course, depend upon how compelling one finds his axioms. There are two questions to be asked: Can his axioms of pure rationality, which express necessary conditions for the existence of the desired expected utility representation, be defended? Can his nonnecessary structure axioms be avoided? Savage's axioms of pure rationality are SAV_2 (Partial Ordering), SAV_4 (Independence), SAV_5 (Nullity), SAV_6 (Stochastic Dominance), and SAV_8^+ (Dominance). Proponents of Savage's version of the thesis of expected utility maximization have no option but to defend these principles; to deny them is to give up the idea that rational preferences must be represented by expected utilities that obey Savage's Equation. Savage's structure axioms – $*SAV_0$ (Act Richness), $*SAV_1$ (Non-triviality), $*SAV_3$ (Completeness), and $*SAV_7$ (Event Richness) – are not strictly necessary for the existence of the representation, but they are needed in Savage's proof of its existence. A proponent of Savage's theory is not committed to defending these axioms as they stand; she can, instead, try to show how their use in the proof is innocuous, or she can argue that they are not to be taken literally (which is what I shall advocate).

The first question, of whether the axioms of pure rationality are true, is not one I am going to focus on here. This is not because I think it unimportant, but because it has been handled so well in other places.[22] I will, instead, concentrate most of my attention on the structure axioms. I have two reasons for doing so. First, as subsequent

---

[22] Let me point the reader to the first few chapters of Patrick Maher's (1993) for an excellent recent treatment of the issues in this area.

discussion will show, a number of objections that have been raised against the rationality axioms are really misdirected attacks against various structure principles. Second, I think that the Achilles' heel of Savage's theory is its dependence on structure axioms that cannot be satisfactorily explained away. This, I believe, ultimately forces proponents of expected utility maximization to look elsewhere for a justification of their theory.

Before considering particular structure axioms, we should take a moment to get clear about the sense in which they are unnecessary for the representation. According to the official definition given at the beginning of this chapter, a probability/utility pair (P, u) *represents* a preference ranking ($>$, $\geq$) over $\boldsymbol{A}$ just in case both

$$A > A^* \text{ only if } \Sigma_S P(S)u(O[A,S]) > \Sigma_S P(S)u(O[A^*,S])$$
$$A \geq A^* \text{ only if } \Sigma_S P(S)u(O[A,S]) \geq \Sigma_S P(S)u(O[A^*,S])$$

hold for every $A, A^* \in \boldsymbol{A}$. Notice that I did not make it a part of my definition that these should be biconditionals, nor that (P, u) should be unique up to a positive linear transformation of u. Some rational choice theorists, Savage among them, use the term *representation* in this stronger sense. They take $U(A) \geq U(A^*)$ to entail $A \geq A^*$, and think that only one P can represent ($.>., .\geq.$) and that U will be uniquely specified once a zero and a unit for measuring utilities have been chosen. This makes the completeness axiom necessary for representability (to get the biconditional). It also forces one to invoke some strong structural condition like *SAV$_7$ to secure the uniqueness of P.[23] For reasons we will be discussing shortly, it is implausible to think that representability in this stronger sense is what instrumental rationality demands. Indeed, it does not seem that rationality should require that a person have any terribly complicated beliefs or desires at all.

Perhaps the clearest way to see this is by asking whether it is irrational to violate the completeness axiom *SAV$_3$. The justifications proposed in the philosophical and economic literature for requiring rational preferences to completely order $\boldsymbol{A}$ seem pretty half-hearted. Kenneth Arrow merely states that "in absence of [Completeness], no choice at all may be possible.[24] R. Duncan Luce and Howard Raiffa justify their version of the axiom by calling it "quite innocuous in the sense that, if a person takes serious issue with [it], then we would

[23] *SAV$_7$ is not required because it is possible for a likelihood ranking to be uniquely represented by a probability even when its base algebra is finite. See Suppes and Zanotti (1976).

[24] Arrow (1966, p. 255).

contend that he is not really attuned to the problem we have in mind."[25] This cavalier attitude can be traced to the old behaviorist idea that an agent prefers one prospect to another just in case he would select the former over the latter if he were forced to choose between the two, with no other options being available. On such an interpretation, it will automatically be true that "preferences" are complete.

As noted in Chapter 1, however, this is a highly implausible view. There are two problems with it that are especially pertinent here. First, the behaviorist view allows preferences to be defined only among prospects that the agent can choose to make the case. Second, it assumes that every choice will reveal a definite preference. The fact is that most of the things about which people have preferences are *not* objects of choice for them. For example, while I would rather not have the gene for Huntington's chorea I fully recognize that this is not a matter over which I have any influence.

Perhaps more importantly, even in cases where a person can make choices among prospects there is no guarantee that her acts will reveal what she desires. One reason is that a single choice can never reveal that she is *indifferent* between two options (since she only gets to choose one or the other). This is not a serious problem for the behaviorist, however, since the most plausible version of behaviorism has it that preferences are *dispositions* to choose rather than choices *per se*. To be indifferent between two prospects on such a view is to be *equally disposed* to select either of them when given the option.[26] Much more serious is the fact that choice can occur even in contexts where the agent has no definite preference at all. The conflation of the lack of preference with indifference is what underlies a number of alleged counterexamples to expected utility theory. Let's consider two. The first has to do with so-called incommensurable goods. As a number of authors have recognized,[27] we evaluate prospects on a variety of "scales" of goodness, and there is no reason, in general, to think that these can be amalgamated in any satisfactory way to yield a single unitary measure of value. Some goods (or ways of being good) are simply *incommensurable* with others. Here is an example that makes the point obvious: You are the dean of admissions at a small

[25] Luce and Raiffa (1957, p. 287).
[26] This disposition might be revealed in one's pattern of preferences in a variety of ways. For example, if "the same" choice is presented to the agent over and over she may end up choosing each option about half the time. More sophisticated devices for eliciting indifference can be found in the literature. See Tversky (1975) for details.
[27] See, for example, Raz (1986, p. 322) and Anderson (1993, p. 55–59).

college that has a chess team, a gymnastic team, and an orchestra. You have the files of three applicants before you. All are similar in every relevant way except the following:

	Gymnastics	Chess	Violin
Applicant 1	Won a gold medal in the most recent Olympic Games	Never can recall how the knight moves	First violin in her local high school orchestra
Applicant 2	Made all-county in high school	Beats Kasparov about two times in five	Can't play a lick
Applicant 3	Complete klutz	Junior chess champ of her small town	Has soloed with all the world's great orchestras

There is one place left in next year's freshman class (this is quite a school you work at!), and you need to decide which of these three candidates will get in. What would you do? It is possible, of course, that you have some reason for weighting one of the categories more heavily than the others (perhaps the gymnastics team already has more gold medalists than it can use), but let's suppose that you do not. Is there a way to decide?

Many people look at this kind of example and see a violation of the transitivity clause of the partial ordering axiom. They reason by the kind of process used in Condorcet voting schemes in which elements of some set of options are judged against one another in pairs. Applicant 1 should be preferred to Applicant 2, they argue, because she scores better on both gymnastics and music and worse only on chess. Applicant 2 must be preferred to Applicant 3 because she scores better in gymnastics and chess. Finally, Applicant 3 should be preferred to Applicant 1 because she scores better on chess and music. *Ergo*, the claim is, it can be rational to have intransitive preferences.

This is obviously a fallacy. Condorcet reasoning is only appropriate when one can be certain that the various criteria carry equal weight, and when one knows that, for example, the difference between being a world class violinist and not playing a lick is too small to offset Applicant 3's deficiencies relative to Applicant 2. You do not know either of these things in this situation! The right thing to say here is not that your preferences are intransitive, but that you lack preferences.

You neither prefer Applicant 1 to Applicant 2, nor Applicant 2 to Applicant 1, nor Applicant 1 to Applicant 3, and so on. You are *not* indifferent among the three candidates; rather, their talents are incommensurable because they meet the standards of goodness for college admission in such different ways. They are all excellent, you might say, but excellent in such diverse ways that there is no way to say who is better. You do have to choose among them, of course, but the choice you make will not indicate that you regard one of the three as better than the other two.

The point here is that the existence of incommensurable goods (or ways of being good) will generally lead to *incompleteness* in an agent's preferences rather than intransitivity. Incompleteness arises when an agent is unable to form any view about the relative value of the outcomes of prospects she considers, or when she lacks sufficient information to form beliefs about the likelihoods that these outcomes have of coming about. In such cases, it is perfectly rational for her not to adopt *any* view about the comparative desirability of the alternatives under consideration. Again, this is not the same as being indifferent. When a person is indifferent between X and Y she makes a definite judgment to the effect that either would contribute equally to her happiness. Incomparability means that she is unable to judge whether X or Y would be the better thing to happen.

One more example might help make the point clearer. Again, it is a case in which a seeming violation of the laws of expected utility maximization is really explained by an incompleteness of preferences. One of the most convincing counterexamples ever offered against the Independence Axiom, SAV_4, is the *Ellsberg Paradox*. Suppose a ball is to be drawn at random from an urn that contains thirty red balls and sixty white or blue balls in unknown proportion. Here are some prospects:

	Red 1/3	White	Blue
		—————2/3—————	
G_1	$100	$0	$0
G_2	$0	$100	$0
H_1	$100	$0	$100
H_2	$0	$100	$100

Decide between G_1 and G_2 and then between H_1 and H_2. What are your preferences? Many people profess to prefer G_1 to G_2 and H_2 to H_1; this violates Independence because H_1 and H_2 only differ from G_1 and G_2 in having $0 in the Blue column replaced by $100. One explanation of this is that people tend to place a value on knowing the objective chances with which their acts will produce various outcomes.[28]

This is probably right in some cases, but I expect that the more common explanation is that people do not really have the preferences that their choices "reveal." They, at least, do not have these as their *considered* preferences. Look back at G_1 and G_2. Are you really convinced that you will be *better off* having G_1 rather than G_2? I personally find it hard to come to any clear judgment when I think about the issue carefully. On one hand, a kind of "balance of reasons" argument seems to apply. Anything to be said in favor of G_1 (e.g., there might be only five white balls in the urn) appears to be offset by something to be said against it (there may be fifty-five white balls in the urn). This would suggest indifference as the right attitude. On the other hand, the idea that the reasons are "in balance" itself seems open to question given my complete ignorance of the relevant chances. Since I do not know that the chance of there being five balls is the same as the chance of there being fifty-five, I do not know whether these reasons really do cancel out. In the end I find myself at sea; I do not really have definite preferences here. I invite readers to reconsider their own preferences to see whether they stand up to scrutiny. It should be kept in mind, however, that the common "preferences" in the Ellsberg problem can falsify Independence only if they are *not* based on a desire to know the objective chances. If an agent prefers G_1 to G_2 or H_2 to H_1 because she has a (noninstrumental) desire for such knowledge, then the problem's outcomes have been underdescribed. Once they are redescribed to take her desire into account (so that, for example, the $100 in G_1 is replaced by "Winning $100 having known all along that one had a 1/3 chance of $100") the commonly observed Ellsberg preferences no longer provide a counterexample to Independence.[29]

Examples like these should convince us that a decision maker can be perfectly rational even when her preferences do not satisfy the completeness axiom *SAV$_3$. There is nothing inherently irrational

---

[28] Brian Skyrms has suggested this. See Skyrms (1984, p. 33).

[29] The thesis that Ellsberg choosers often do not have definite preferences has been defended by Jeffrey in his (1987) and by Levi in his (1986). It should also be noted that Patrick Maher and Yoshihisa Kashima have carried out a series of studies that show that experimental subjects who have the common pattern of preferences in the Ellsberg problem tend to offer obviously incoherent justifications for their attitudes. See Maher (1993, pp. 68–70). I take this incoherence as evidence for the claim that the "preferences" these subjects "exhibit" are, at the very least, not well thought out.

about not being able to make a definite judgment about which of two prospects is more desirable. Recognition of this point has led many proponents of expected utility theory to adopt an interpretation of structure conditions like *SAV$_3$ that treats them as *requirements of coherent extendibility* rather than strict laws of rationality.[30] A rational agent's preference ranking need not satisfy the structure axioms outright, the story goes, but it should always be possible to extend the ranking to one that does satisfy them without violating any axiom of pure rationality in the process. For example, it is rational for a ranking ($>$, $\geq$) to violate *SAV$_3$ provided that it has a complete extension ($>^+$, $\geq^+$) to all of **A** that obeys *SAV$_3$ and all of Savage's other axioms. To see how preferences can run afoul of this criterion, consider a ranking that contains $(A_E \ \& \ C_{\neg E}) > A$, $A > (A_E \ \& \ B_{\neg E})$, $B \geq C$, and $C > (B_E \ \& \ C_{\neg E})$. While this does not violate any axiom explicitly, it is impossible to extend it in a way that does not violate either Partial Ordering or Independence. (It is an instructive exercise to work out why.)[31]

As Richard Jeffrey has observed, one useful way to think about the extendibility interpretation of structure axioms is through an analogy with Lindenbaum's Lemma from deductive logic.[32] This lemma provides a way of characterizing consistent collections of sentences in terms of extendibility. It says that a set of sentences in a first-order language is consistent exactly if it can be extended to a maximally consistent set, that is, iff one can continue adding sentences to it, without generating any contradiction, until one arrives at a set in which every sentence or its negation appears. The same thing is supposed to happen here. Think of the axioms of pure rationality as the "laws of logic" in this case, and of a contradiction in preference as a situation in which $A > B$ and $B \geq A$. The extendibility interpretation of structure axioms then says that a system of preferences is rational just in case it can be extended to a complete set without violating any "law of logic."

[30] See Kaplan (1983, p. 558) and Jeffrey (1983, pp. 138–41).

[31] The price of the extendibility interpretation is paid in uniqueness. Savage's theorem yields a representation in which P is unique and u is unique up to the arbitrary choice of a zero point and unit to measure utility. However, when its structure axioms are interpreted as conditions of coherent extendibility likelihood and preference rankings will have many representations. This is because a ranking that does not obey all the structure axioms needed to establish a representation theorem will usually extend in more than one way to rankings that do satisfy these axioms. Each of the extended rankings will have its own probability/utility representation, and all of these will represent the initial ranking as well.

[32] Jeffrey (1983). Jeffrey actually makes this analogy in the context of a discussion of partial belief, but it works just as well here where preference is the issue.

While there is something to this analogy, it is not perfect. There are two points of contrast between the cases. First, in deductive logic we have another, more useful, way of characterizing consistency. The completeness theorem tells us that a set of sentences is consistent if and only if its closure under standard rules of deductive inference contains no explicit contradictions. We know, in other words, that our system of rules of inference is sufficiently complete to flush out any inconsistency that may be hiding in a set of sentences. This makes it possible for us to say what consistency requires *without* having to speak about extendibility; consistency requires nothing more or less than obedience to the laws of first-order logic. Think of how odd deductive logic would look if we could not do this. Suppose, for instance, that logicians had discovered *modus ponens*, De Morgan's Laws, and the Law of Noncontradiction, but not the rule of Conditional Proof, and thus had only isolated an incomplete deductive system. Imagine that, in spite of this, they had somehow derived Lindenbaum's Lemma. Their advice to us would then take the following strange form: Make sure that the set of sentences you hold true does not generate an explicit contradiction when you apply *Modus Ponens* and De Morgan's Laws to it, and then make sure that you can extend it to a complete set without violating the Law of Noncontradiction. This last recommendation leaves a lot to be desired, however, since it is almost useless except in certain simple cases where one can see directly that no extension is possible. One would still need to know what it takes, at the "local" level, to make sure that the extendibility condition is satisfied. In an ideal deductive logic, the appeal to extendibility should ultimately be dispensable in favor of a complete set of rules of inference.

The odd situation just described is very much like the one that an extendibility interpretation of the structure axioms puts us in. No one has yet discovered a system of necessary and sufficient conditions for the representability of a preference ranking by an expected utility function. Every existing representation theorem employs at least some structure axioms. Thus, a proponent of expected utility maximization cannot simply give a decision maker a list of axioms of pure rationality, tell her to obey them, and be done with it. Some sort of appeal to the possibility of coherent extendibility seems to be required. This is not an objection to the expected utility hypothesis *per se*, but it does indicate that there is still work to be done. The extendibility interpretation of structure axioms is good as far as it goes, but it is at most a station on the way to a "completeness theorem" that would provide necessary and sufficient conditions for representability.

Another disanalogy between the decision theoretic and logical cases has to do with the richness of the set of objects over which the extension is to be made. Lindenbaum's Lemma does not require incomplete consistent sets to be extendible to highly structured complete sets. (In fact, the completion need not contain any sentences whose nonlogical terms do not figure explicitly in the sentences of the incomplete set.) This is not true when we interpret Savage's structure axioms as the extendibility conditions. The two "richness conditions," *SAV$_0$ and *SAV$_7$, make the set of "acts" **A** over which preferences are defined into a large and complex object. It is unreasonable to require an agent to have preferences that can be extended to a complete ordering of such a rich set of prospects.

*SAV$_0$ and *SAV$_7$ are not on a par here. The latter is included in the theory to assure that the agent's likelihood ranking is atomless. It is not objectionable as an extendibility condition because, as it turns out, almost any likelihood ranking that can be represented by a probability function can be represented by a probability function that is defined over an atomless algebra of propositions. The relevant mathematical fact is

Theorem 3.4. *Let P be a countably additive probability defined on a σ-algebra of propositions Ω. If P does not assign positive probability to any single possible world, then there is a σ-algebra Ω$^+$ that contains Ω as a subset and a countably additive probability P$^+$ defined on Ω$^+$ that both extends P and is atomless in the sense that any proposition X ∈ Ω$^+$ such that P$^+$(X) > 0 can be partitioned into disjoint parts X$_1$, X$_2$ ∈ Ω$^+$ such that P$^+$(X$_1$) > 0 and P$^+$(X$_2$) > 0.*

Proof Sketch (which can be easily filled in by those who know a little measure theory and can be safely skipped by those who do not): We define Ω$^+$ and P$^+$ recursively. Let Ω$_0$ = Ω and P$_0$ = P. At each stage j, define Ω$_{j+1}$ by taking each atom X ∈ Ω$_j$ such that P$_j$(X) > 0 and splitting it into two disjoint parts X$_1$ and X$_2$ each of which contains uncountably many possible worlds. These propositions will not be in Ω$_j$, of course, but we know we can always find them because (i) no single possible world is ever assigned positive probability by P$_0$, which means that each proposition of positive P$_0$ probability must be entailed by uncountably many possible worlds, (ii) each stage of the construction just described preserves (i) for P$_0$, P$_1$, P$_2$,.... Set P$_{j+1}$(X$_1$) = P$_{j+1}$(X$_2$) = P$_j$(X)/2, and define P$_{j+1}$ as the unique finitely additive

105

probability that extends P_j to Ω_{j+1} subject to these constraints. In the limit, define $\Omega^+ = U_j\Omega_j$, and let $P^+(X) = P_j(X)$ if $P_j(X)$ is defined for some P_j (and hence for any P_k with $j < k$). P^+ is obviously finitely additive since each P_j is finitely additive. One can apply the Caratheodory Extension Lemma to get countable additivity. ∎

It follows from this that, except in the exceptional case when an agent assigns a positive subjective probability to some single possible world, any likelihood ranking $(.>., .\geq.)$ that can be represented by a probability P will be representable by a probability P^+ like the one just described. The ordinal likelihood ranking defined by

$$X .\geq.^+ Y \text{ if and only if } P^+(X) \geq P^+(Y)$$

will then be a complete, atomless extension of $(.>., .\geq.)$. Thus, probabilistic representability can *only* be attained for comparative likelihood rankings that can be extended to complete, atomless rankings. This makes it reasonable to view $*SAV_7$ as an extendibility condition.

The other richness condition, $*SAV_0$, cannot be understood in this way. This axiom requires the existence of constant acts that produce the same outcomes in every state of the world, and it requires the set of acts to be closed under "mixing," so that $(A_E \, \& \, B_{\neg E})$ is in **A** whenever A and B are in **A** and E is in $\Omega(\mathbf{S})$. One can interpret the latter clause as requirement of coherent extendibility. If a person's *unconditional* preferences cannot be coherently extended to a complete ranking on a set of prospects that is closed under mixing relative to an event E, then there will be no set of preferences-conditional-on-E that she can rationally adopt. To see why, suppose her preferences can be extended to a complete ranking on some family of prospects **K** that is *not* closed under mixing, but that it can*not* be coherently extended to the set $\mathbf{K}_E = \{A_E \, \& \, B_{\neg E}: A, B \in \mathbf{K}\}$. This means that every extension $(>^+, \geq^+)$ of the person's preferences to $\mathbf{K}_E$ will contain both $(A_E \, \& \, B_{\neg E}) >^+ (C_E \, \& \, B_{\neg E})$ and $(C_E \, \& \, B_{\neg E}) \geq^+ (A_E \, \& \, B_{\neg E})$ for some A, B, C in **K**. Her unconditional desires for prospects in **K** thus make it impossible for her to adopt any coherent view about the desirabilities of those same prospects on the condition that E is true. This clearly is a flaw that expected utility theory cannot tolerate. It is only possible to represent an agent's unconditional preferences by an expected utility if it is also possible to represent her preferences conditional-on-E by an expected utility. As the following result indicates, the existence of a representation for $(>, \geq)$ always determines a representation for $(>_E, \geq_E)$ via a simple rule.

Theorem 3.5. *If (P, u) represents (>, ≥) and E is nonnull, then (P_E, u) represents $(>_E, ≥_E)$ where P_E is the ordinary conditional probability $P_E(X) = P(X \& E)/P(E)$.*

Proof: If (P, u) represents (>, ≥), then for any acts $A, C \in \mathbf{A}$ one has $A >_E C$ if and only if $(A_E \& B_{\neg E}) > (C_E \& B_{\neg E})$ for some $B \in \mathbf{A}$. It follows directly that

$$A >_E C \text{ only if } \Sigma_S P(S \& E)u(O[A,S]) + \Sigma_S P(S \& \neg E)u(O[B,S])$$
$$> \Sigma_S P(S \& E)u(O[C,S]) + \Sigma_S P(S \& \neg E)u(O[B,S])$$

But, this latter inequality holds exactly if

$$\Sigma_S P(S \& E)u(O[A,S]) > \Sigma_S P(S \& E)u(O[C,S]).$$

The desired result then follows when we multiply through by $1/P(E)$. The case for weak preference is similar. ∎

Theorem 3.5 makes it reasonable, indeed mandatory, for proponents of Savage's theory to regard the second clause of $*SAV_0$ as a requirement of coherent extendibility.

Not so the first. This part of the axiom says that if it is possible for a decision maker to attain a given level of overall satisfaction in one state of the world, then it is also possible for her to attain that same level of satisfaction in any other state of the world. Indeed, it says something even stronger: namely, that for any outcome O there will always be an act in $\mathbf{A}$ that produces O *come what may*. To illustrate just how strong a requirement this is, imagine some exceptionally desirable future course of events O – world peace breaks out, cures for all known diseases are found, your children marry well, you win the Nobel Prize, or whatever else suits your fancy. Now, let S be the state in which all living things on earth are destroyed at noon tomorrow by the impact of a huge asteroid. If $*SAV_0$ is correct, then there must be some course of future events that is consistent with S and that would be just as good, from your perspective, as the wonderful outcome O. In fact, there has to be some contingent proposition A_O whose truth will guarantee you this wonderful outcome if S, or anything else, occurs. Savage calls A_O an act, of course, but we have already seen that it cannot be identified with something anyone might *do*. However, this is only part of the trouble. The real problem is that A_O might describe an *impossible* state of affairs. It would not be at all irrational for you to have desires that make it impossible for anything that happens before noon tomorrow to compensate you for the asteroid's devastat-

107

ing effects.[33] If your desires are like this, then there will be nothing that can perform A_O's function: There will be *no* proposition that is consistent with both the destruction of the world and all the wonderful things that O promises whose truth would make you indifferent between the two prospects. Under such conditions, it would surely be unreasonable for a decision theorist to demand that your preferences be extendible to a complete ranking defined over a set that contains A_O. It clearly should not be a requirement of instrumental rationality that you be able to adopt coherent preferences about prospects that your preferences themselves render impossible.

In my view, the reliance on constant acts is a serious and irremediable flaw in Savage's theory. Since the proof of his representation theorem presupposes the existence of a constant act for each outcome, and since this presupposition cannot be explained away as an implicit extendibility requirement, I think we must recognize that Savage ultimately fails to provide a satisfactory justification for the expected utility hypothesis. I thus find myself in agreement with Duncan Luce and Patrick Suppes, who have written that

the most serious weakness of Savage's system, and all those similar to it, is the essential use of the constant [acts], that is, those [acts] that have a constant outcome independent of the state of nature. In actual practice it is rare indeed for such [acts] to be realizable; each outcome cannot be realized for some possible state of nature. Unfortunately, there seems to be no simple method of getting around their use.[34]

The root of Savage's problem can be traced to his assumption that *all* functions from states into outcomes should be included in the set of acts. Many of these functions, like the constant acts, simply do not correspond to any possible states of affairs.

Duncan Luce and David Krantz have reformulated Savage's Theorem in a way that does not require constant acts.[35] They succeed in doing this by taking the decision maker's preferences to be defined over *partial* (as well as total) functions from states to outcomes. These correspond to the propositions we have been calling *conditional actions*. (Luce and Krantz refer to them as *conditional decisions*.) They prove a general representation result that shows how a pre-

---

[33] People sometimes reply by pointing out that the destruction of life on earth might not be so bad if there were a god, and if one ended up in heaven after the disaster. True enough, but the basic problem remains since there is nothing preventing us from rewriting the example so that S includes the statement that there is no god and that the asteroid destroys all sentient beings.

[34] Luce and Suppes (1965, p. 299).

[35] Luce and Krantz (1971).

ference ranking over conditional acts that obeys certain axioms can be represented by a probability utility pair (P, u) in such a way that

$$A_E > B_E \text{ only if } U(A_E) > U(B_E)$$
$$A_E \geq B_E \text{ only if } U(A_E) \geq U(B_E)$$

where

(L & C) $\quad U(A_E) = \Sigma_S P_E(S) u(O[A, S]) = \Sigma_j P_E(E_j) U(A_{Ej})$

where, again, P_E is the standard conditional probability $P_E(X) = P(X \& E)/P(E)$, and where the E_j in the rightmost expression may be any finite partition of E from $\Omega(\mathbf{S})$. Notice that Savage's theorem is a special case of this when E is the necessary proposition **T**.

I will not list all of Luce and Krantz's axioms here, but I do want to call attention to three of them that lead to trouble. The first two are closure conditions on the set $\mathbf{A}^C$ of all conditional acts. The third is a richness condition.

- **Closure under Restrictions.** If E and F are events in $\Omega(\mathbf{S})$ and if F entails E and F is not null, then A_E is in $\mathbf{A}^C$ only if A_F is also in $\mathbf{A}^C$.
- **Closure under Conjunctions.** If E and F are mutually incompatible events in $\Omega(\mathbf{S})$, and if both A_E and B_F are in $\mathbf{A}^C$, then $A_E \& B_F$ is also in $\mathbf{A}^C$.
- **Richness.** Given any $A_E \in \mathbf{A}^C$ and any nonnull $F \in \Omega(\mathbf{S})$ there exists a $B_F \in \mathbf{A}^C$ such that the decision maker is indifferent between A_E and B_F.

The first condition stipulates that any conditional act defined over E is also defined over any subevent of E that has any chance of occurring. The second is a generalization of the "closure under mixing" portion of *SAV$_0$. The third is a strong requirement to the effect that no two events can be so far apart, from the perspective of the decision maker's desires, that the differences between them cannot be effectively canceled out by an appropriate choice of "acts."

Luce and Krantz are able to exhibit a decision problem that satisfies all their axioms and yet contains no constant acts, and they regard the fact that their theory can do without constant acts as one of its major advantages over Savage's account.[36] This, however, is a Pyrrhic victory. Even without the use of constant acts, the Luce/Krantz theory requires decision makers to have preferences regarding prospects that may well be impossible. One good example is what I call a *mitigator*. A mitigator is a conditional act that counteracts the effects of events

[36] Luce and Krantz (1971, pp. 261–62).

that one would expect to generate rather different sorts of outcomes. A mitigator for a pair of incompatible events E and F is a conditional act $A_{E\vee F}$ such that the decision maker regards A_E and A_F as equally desirable. Knowing that such a mitigating proposition is true makes the agent entirely apathetic about the question of whether E or F occurs. To illustrate, suppose you hold a wager in which you stand to win \$1,000 if a coin lands heads (E) and to lose \$1,000 if it lands tails (F). A mitigator would be a circumstance that would make you indifferent to the result of the coin toss, for example, being covered by an insurance policy that pays \$2,000 in the event of a tail. To prove their result, Luce and Krantz must assume that mitigators of this type exist, not merely for some choices of E and F, but for all. (This is a consequence of the three axioms given.)

Mitigating acts are only a small improvement over Savage's constant acts (which function as mitigators in his theory). Consider again an event E in which all kinds of good things happen and the event F in which the earth is destroyed by an asteroid. While Luce and Krantz do not require that there should be an act that generates *the same* outcome in both these cases, they do require that there be an act whose results the decision maker expects to be *equally desirable* in either event. Indeed, for any act A that is compatible with E, no matter how wonderful its outcomes, there must be an act B compatible with F that offsets the deleterious effects of the asteroid to such an extent that the agent is entirely indifferent between E and F given that A_E & B_F holds. As with Savage's constant acts, it is not clear why we should regard acts such as B, or conditional acts such as B_E, as real possibilities. Given this, there is no reason to require a rational agent's preference ranking to be extendible to a family of prospects containing such fantastic elements.[37] In the end, I do not see how it is possible to formulate anything like Savage's theory without resorting to mitigators. This gives us one good reason to look elsewhere for a representation theorem for expected utility theory.

3.4 THE PROBLEM OF SMALL WORLDS IN SAVAGE'S THEORY

A further reason to seek an alternative foundation for expected utility theory is supplied by the problem of "small worlds" that was discussed

[37] Another theory that suffers from this difficulty is found in Fishburn (1973, 1974). Fishburn's account differs from that of Luce and Krantz in a number of respects (e.g., he assumes that the set of acts is closed under mixtures), but he too must postulate mitigating acts. He recognizes this as a problem: "The only way to salvage our system is to introduce some artificial prizes or penalties, or artificial acts, that will rectify the difficulty" (1974, p. 27).

at the end of the previous chapter. The challenge, recall, was to formulate a cogent account of instrumental rationality for small-world decision problem making by (a) explaining what it takes for an agent contemplating a small-world decision to correctly estimate her grand-world attitudes toward its outcomes, states, and acts, and (b) formulating a unified account of rationality that applies to both grand- and small-world decisions. Such a unified account must guarantee that an agent who correctly estimates her grand-world attitudes, and who adheres to the laws of rationality, will make a choice that is rational whether it is evaluated from the grand-world or the small-world perspective. To see how this problem arises in the context of Savage's theory,[38] let $\boldsymbol{D}^G = (\Omega^G, \boldsymbol{O}^G, \boldsymbol{S}^G, \boldsymbol{A}^G)$ be an agent's grand-world decision problem and $\boldsymbol{D} = (\Omega, \boldsymbol{O}, \boldsymbol{S}, \boldsymbol{A})$ be some small-world version of it (so that the actions, states, and outcomes in $\boldsymbol{D}$ are, respectively, disjunctions of actions, states, and outcomes in $\boldsymbol{D}^G$). Savage explicitly developed his theory with the grand-world context in mind. His basic criterion of rationality requires that an agent's preference ranking over *grand-world* acts be represented by some expected utility of the form $U(A) = \Sigma_S P(S)u(O[A, S])$ where A, S, and $O[A, S]$ denote grand-world acts, states, and outcomes, respectively. Savage knew, however, that his system would not be entirely adequate unless it could be applied to both grand- and small-world decision problems. In *The Foundations of Statistics* he frankly admitted that "any claim to realism made by this book – or indeed by almost any theory of personal decision of which I know – is predicated on the idea that some of the individual decisions into which people tend to subdivide the single grand decision do replicate in microcosm the mechanism of the single grand decision."[39] Savage thus recognizes that his theory cannot pass the "realism" test unless it is (at least sometimes) the case that rationality requires that an agent's preference ranking over *small-world acts* be represented by an expected utility of form $U^*(B) = \Sigma_E P^*(E)u^*(O[B, E])$ where B, E and $O[B, E]$ range over *small-world* acts, states, and outcomes. In both the grand- and small-world cases, then, Savage is committed to the view that instrumental rationality requires

---

[38] My discussion of the problem of small worlds differs somewhat from Savage's owing to my formulation of his theory in propositional terms and my somewhat different characterization of decision problems. I do not think these differences affect the basic point at issue. Moreover, insofar as they do, I am convinced that my treatment of the problem is superior. For example, Savage identifies small-world outcomes with grand-world acts. This is an artifact of his decision to treat acts as functions from states to outcomes, and it greatly restricts the scope of his approach.

[39] Savage (1954/1972, p. 83).

representability by some utility/probability pair that obeys his equation.

To ensure the existence of such representations Savage stipulates that an agent's preference ranking over both small- and grand-world acts must satisfy his postulates.[40] This is problematic in itself because it requires small-world decisions to have a great deal of formal structure (constant acts, conditional acts, infinitely many states, and so on), which makes them far more complicated than any actual decision could ever be. Even more serious is the fact that mere obedience to Savage's axioms cannot ensure that the probability and utility functions that represent the decision maker's small-world attitudes will agree with the probability and utility functions that represent her grand-world attitudes. If the agent's preference rankings over A and A^G both satisfy the axioms, then Savage's Representation Theorem ensures that there will be probability/utility pairs (P*, u*) and (P, u) such that the agent's preferences over A are represented by (P*, u*) and her preferences over A^G are represented by (P, u). What is *not* guaranteed, however, is that these probabilities and utilities will agree with one another in any meaningful way. Both $P(E)$ and $P*(E)$ will be defined for any small-world state (grand-world event) E, but there is no reason to think that these probabilities will be equal. Likewise, nothing forces the u* utility of a small-world outcome to bear any specific relation to the u utilities of the grand-world outcomes consistent with it. It does not follow, for example, that $u*(O[B, E])$ is a probability weighted average of the values $u(O[A, S])$ where A entails B and S entails E. Finally, even if we do assume some definite mathematical relationship among the utilities of small- and grand-world outcomes, their associated expected utilities for acts will not necessarily be consistent since, in virtue of the potential disparity between P and P*, it might happen that $U*(B_1) > U*(B_2)$ even though $U(A_2) \geq U(A_1)$ for every pair of grand-world acts such that A_1 entails B_1 and A_2 entails B_2. The general problem, then, is that it is possible for a decision maker's preference rankings for the acts in D and in D^G to each be represented in exactly the way that Savage requires even when there is no *joint* probability/utility representation that encompasses both rankings. The agent is thus afforded no assurance that the policy of conforming her preferences to Savage's axioms in small-world contexts will lead her to make choices that would agree with the grand-world evaluations she might make if she had time to consider

---

[40] Some postulates (SAV$_1$ and SAV$_2$ for example) are satisfied automatically in the small world if they are satisfied in the grand world. Others (like *SAV$_7$) must be imposed anew.

her predicament in complete detail. Since a solution to the problem of small worlds involves giving such an assurance, it follows that Savage's theory cannot solve it.

Savage recognized this as a serious deficiency in his account, and it bothered him greatly that he could not remove it. He expressed the hope that there would be only a very "remote possibility"[41] of conflict between small- and grand-world representations, but this hope was never backed up by any rigorous proof. There was a good reason for his failure. As we will see in the next chapter, Savage's system lacks a crucial resource required to solve the problem of small worlds: namely, a way to make sense of the possibility that the probabilities of states of the world might vary, depending on what action the agent chooses to perform. Richard Jeffrey has developed a decision theory that takes this possibility into account. It is, as we shall discover, the only theory that does not fall victim to the problem of small worlds.

---

[41] Savage (1954/1972, p. 90).

4

Evidential Decision Theory

In this chapter we examine the *evidential* decision theory developed by Richard Jeffrey in *The Logic of Decision*. Jeffrey's theory has a number of significant advantages over Savage's. For example, it is able to account for the effects of an agent's actions on the probabilities of states of the world, and it provides a neat solution to the problem of small worlds. Even more important is the fact that the theory can be underwritten by a beautiful representation result, proved by the mathematician Ethan Bolker, that has almost none of the defects associated with Savage's theorem. This theorem does not presuppose the existence of either "constant acts" or "mitigators," and all of its structure axioms can be reasonably interpreted as extendibility conditions.

We will learn in the next chapter that evidential decision theory cannot be used as a theory of rational *choice* since there are cases in which its prescriptions tell an agent to perform acts that are sure to leave her worse off come what may. This does not mean, however, that Jeffrey's theory is irrelevant to questions about what people should do. Indeed, I shall argue that any adequate account of rational action must be based on an underlying theory of *valuing* that has precisely the form Jeffrey proposes. The error in evidential decision theory, it turns out, is not found in the constraints it imposes on rational desire, but in a mistaken assumption about the *epistemic standpoint* from which decision makers should evaluate their acts. Evidential decision theory will thus be seen to be an essential element in our understanding of rational desire and rational action even though it does not have precisely the role that Jeffrey thought it did.

Evidential decision theory does have one peculiar feature, however. It turns out that the constraints it imposes on rational preference are insufficient to fix the canons of rational belief. Indeed, the axioms used in proving Bolker's theorem can be satisfied by an agent whose comparative estimates of likelihood cannot be represented by *any* probability function. I will argue that this is an advantage, not a drawback, of the theory because it helps us to recognize that a complete account of rationality must supplement constraints on desires with independently motivated constraints on beliefs. In this spirit, we

114

will see how to extend Bolker's theorem by augmenting it with axioms pertaining to judgments of comparative likelihood, and then showing how this total set of requirements on desires and beliefs can be used to explain and justify the policy of expected utility maximization. The resulting representation theorem will serve as the basis for further developments.

4.1 THE NEED FOR EVIDENTIAL DECISION THEORY

Savage's prescription to maximize expected utility defined by his equation only makes sense under the assumption that a rational agent's beliefs about the likelihoods of states of the world do not vary depending upon which act she performs. This assumption is implicit in both the "Independence Axiom" SAV_4 and the "Nullity" axiom SAV_5. It comes out most clearly when we examine the following important consequence of these principles:

Dominance. Let E and F be mutually incompatible events. If A is weakly preferred to B both given E and given F, then A is also preferred to B given $E \lor F$. Moreover, if either initial preference is strict and its associated event is nonnull, then the final preference is strict as well. In mathematese:

$$A \geq_E B \text{ and } A \geq_F B \text{ imply } A \geq_{(E \lor F)} B \text{ and } A >_E B \text{ and } A \geq_F B$$
$$\text{imply } A >_{(E \lor F)} B \text{ as long as } E \text{ is not null}$$

This is a formal expression of the sort of "sure thing" reasoning that Savage's home buyer employed in the passage quoted in the previous chapter. In that instance, as in many others, Dominance recommends the right course of action. Indeed, it is hard to imagine any principle of rational decision making that has such a wide range of correct applicability.

There are, however, cases where it leads us astray. As Jeffrey recognized,[1] preferring the dominant option can lead to unwise decisions when the probability of events depends on one's choice of an action. Consider the following "deterrence" example: Suppose you have just parked in a seedy neighborhood when a man approaches and offers to "protect" your car from harm for $10. You recognize this as extortion and have heard that people who refuse "protection" invariably return to find their windshields smashed. Those who pay find their cars intact. You cannot park anywhere else because you are late for an

[1] See Jeffrey (1965/1983, pp. 8–9).

important meeting. It costs $400 to replace a windshield. Should you buy "protection"? Dominance says that you should not. Since you would rather have the extra $10 both in the event that your windshield is smashed and in the event that it is not, Dominance tells you not to pay. Of course, this is absurd. Your choice has a direct influence on the state of the world; refusing to pay makes it likely that your windshield will be smashed while paying makes this unlikely. The extortionist is a despicable person, but he has you over a barrel and investing a mere $10 now saves $400 down the line. You should pay now (and alert the police later).

When confronted with this sort of case, Savage responded by suggesting that his theory had been misapplied.[2] The trouble is not caused by Dominance *per se*, he claimed, but by its application to an ill-posed problem. It is only legitimate to invoke the principle, or to use the Savage formalism generally, in a decision problem whose states have been chosen to ensure that the agent's beliefs about them will not vary depending upon which actions she performs. States, in other words, must be probabilistically independent of acts if the prescription to maximize unconditional utility is to make sense. So, for example, the appropriate framework for the deterrence problem described here is not one based on the state partition

E: The windshield will be broken
$\neg E$:   The windshield will not be broken

but on the partition

E_1:   The windshield will be broken whatever you do
E_2:   The windshield will not be broken if you pay, but will be broken if you do not pay
E_3:   The windshield will be broken if you pay, but will not be broken if you do not pay
E_4:   The windshield will not be broken whatever you do

Then, since people generally get their windshield broken if and only if they refuse to pay, it follows that E_2 is likely, while E_1, E_3, and E_4 are not. E_2 is thus going to rule in any calculation of expected utility, which means that Savage's theory does give the correct answer after all: You should pay the extortionist.

This suggestion is good as far as it goes, but it leaves two questions unanswered. First, the notion of a state's probability being "independent" of an action is ambiguous between a *statistical* or *evidential* reading and a *causal* or *subjunctive* reading. On the first interpretation

---

[2] My discussion here is guided by Jeffrey (1965/1983, pp. 21–22).

an event is independent of an act just in case the agent would not take her performance of the act to provide *evidence* for thinking that the event will occur or that it will not occur. This sort of *evidential independence* is clearly lacking in the initial version of the deterrence case; by refusing to pay the hooligan $10 you give yourself strong evidence for thinking that he will break your windshield. But when the events are described as $E_1, \ldots, E_4$ this evidential connection is broken. On a causal interpretation, an event is independent of an act just in case performing the act does nothing to *bring about* the event or to *prevent* it. Here again, we have an illegitimate act-to-state dependence in the first version of the deterrence problem – not paying the extortionist would be one of the causal factors leading up to the breaking of your windshield – but we do not have it in the second. Despite the fact that evidential and causal independence go together in this case (and many others), we will see that this is not inevitable. Thus, there is an ambiguity lurking in Savage's treatment of deterrence problems. Should states in well-formed decisions be independent of acts in the causal sense or in the evidential sense?

Even if Savage had given us an unambiguous answer to this question, there would be a still more pressing concern about his handling of this case. The issue of whether the states used in framing a decision problem are independent of actions is one that can only be decided by an appeal to the decision maker's opinions, in particular, to her opinions about propositions that involve both states and acts. In the case of evidential independence, for example, the correct criterion is this:

> S is evidentially independent of A exactly if the agent's degree of confidence in S on the condition that A is true is exactly as strong as her degree of confidence in S on the condition that A is false.

The correct criterion for the kind of causal independence needed in decision theory is (to a first approximation)

> S is causally independent of A just in case the agent believes the subjunctive conditional $A \;\square\!\!\rightarrow S$ exactly as strongly as she believes $\neg A \;\square\!\!\rightarrow S$.

Unfortunately, there is no way to express either of these ideas within the Savage framework because it has no place for beliefs that "mix" events and actions. For Savage, subjective probabilities attach only to events, and this makes it impossible within the theory to specify the conditions that an agent's preferences would have to meet in order for his axioms to apply to her predicament.

117

This would not be a major problem if there were some canonical way of constructing state and act partitions that would be independent of one another relative to *any* subjective probability function. The preceding transformation of E and $\neg E$ into $E_1, \ldots, E_4$ may even suggest such a construction. When a decision problem has "act-dependent" states, one might think, one can always replace its partition of states **S** by a new partition **S*** that contains all consistent propositions of the form $S^* = \&_A(A \Rightarrow S_A)$ where A ranges over elements of **A** and each S_A is in **S**. The claim would then be that this new state partition **S*** will always contain states that are independent of actions relative to any reasonable subjective probability function.

There are two reasons why this maneuver will not work. First, it does not sit comfortably within Savage's theory since it forces us to rewrite the act partition so that acts have the form $\&_S(S^* \Rightarrow O[A, S^*])$, and it is hard to make any intuitive sense of such propositions (write one out and try). Second, and more importantly, it is not clear that the trick works. Consider $E_1, \ldots, E_4$. Must these events really be independent of what you do relative to *any* reasonable set of beliefs you might have? It does not seem so. Let's add a bit to the story: Suppose that you think that the hooligan is still trying to decide what *he* is going to do, so that none of $E_1, \ldots, E_4$ is yet carved in stone. In such a circumstance you might think that giving him the money quickly will make it likely that he will decide on the policy described in E_2, while giving it to him slowly will make it likely that he will choose E_1. In this case you will *not* see the E_j as being independent of your action. Of course, one might try the canonical trick again at this point, and say that the correct states are things like

> *F*: If you pay, then the hooligan will adopt the policy of breaking your windshield if and only if you pay quickly, and if you do not pay, then the hooligan will adopt the policy of not breaking your windshield whatever you do

Even if one does this, however, there is still no guarantee that one will end up with a set of states that a rational agent will necessarily regard as independent of her acts. In the end, the point is a logical one. In Savage's theory **S** and **A** are always going to be logically independent partitions, so that $(A \& S)$, $(A \& \neg S)$, $(\neg A \& S)$, and $(\neg A \& \neg S)$ will all be consistent for each A and S, and it is *always* possible to find *some* probability distribution over these propositions that makes A relevant to S.

The basic problem for Savage's theory is that its equation for expected utility defines the utility of every act A in terms of *the same* probability distribution over events, so that the values of P(E) used to compute U(A) do not take A's truth into account. To handle deterrence cases properly, one needs to find a characterization of the conditions under which an agent regards S as independent of A in either the evidential or the causal sense. Jeffrey accomplished this for the evidential case.

4.2 JEFFREY'S THEORY

Jeffrey solved the deterrence problem without placing "external" restrictions on the types of decision problems to which expected utility theory can be applied by replacing Savage's formula for unconditional expected utility by a new, *conditional* utility whose defining equation is[3]

Jeffrey's Equation. $V(A) = \sum_S P(S/A) u(O[A, S])$
$$= \sum_S P(S/A) u(A \ \& \ S)$$

Here P($\bullet$/A), also written $P_A(\bullet)$, is the usual conditional probability distribution over events defined by P(E/A) = P(E & A)/P(A) whenever $E \in \Omega(\mathbf{S})$ and P(A) > 0. The quantity V(A) is A's *evidential expected utility*. Jeffrey's proposal is that rational decision makers should always align their preferences by increasing *evidential* expected utility and should always act so as to maximize V(A).

To see how Jeffrey's theory differs from Savage's it is useful to look at the *difference* between Jeffrey's V(A) and Savage's U(A)

$$V(A) - U(A) = \sum_S [P(S/A) - P(S)] u(O[A, S])$$

Focus on the term P(S/A) − P(S). On a Bayesian reading, this is one way of measuring the degree to which A is *evidentially relevant* to S. When [P(S/A) − P(S)] > 0, A's truth provides evidence in favor of S, and the agent's degree of confidence in S should increase if she learns A. When [P(S/A) − P(S)] > 0, A is evidence against S, and the agent's degree of confidence in S should decrease as a result of learning A. When P(S/A) = P(S), A and S are *statistically independent*, and learning A should not alter the agent's confidence in S. Jeffrey's definition of utility assigns a greater weight to the outcome O[A, S] in precisely those cases in which A is positively evidentially relevant to S. O[A, S]

---

[3] Recall that propositions of the form A & S are the *atoms* of Ω, so that every element of the algebra can be expressed as a disjunction of such propositions.

is weighted less heavily when A is negatively relevant to S. The weights are equal when A and S are statistically independent. On balance then, $V(A)$ is a measure of the extent to which learning A's truth would provide the decision maker with *evidence* for thinking that desirable outcomes will ensue. Jeffrey makes this point by referring to $V(A)$ as A's "news value" because it measures the extent to which the decision maker would welcome news that A is true. $V(A)$ thus provides a quantitative representation of her judgments about the *auspiciousness* of various prospects, that is, of their values as *indications* or *signs* of desirable or undesirable results.

The most important formal difference between Jeffrey's theory and Savage's is that the former makes it possible to assign expected utilities to propositions other than those describing actions. Indeed, for any decision problem $\boldsymbol{D} = (\Omega, \boldsymbol{O}, \boldsymbol{S}, \boldsymbol{A})$, Jeffrey's Equation allows us to assign a news value to *every* element of the algebra Ω. Given a subjective probability defined on Ω and an assignment of unconditional utilities to elements of $\boldsymbol{O}$ or, equivalently, to all conjunctions of the form $A \& S$ with $A \in \boldsymbol{A}$ and $S \in \boldsymbol{S}$, it makes sense to write the news value of any proposition $X \in \Omega$ (which, recall, can always be expressed as a disjunction of act/state conjunctions) as[4]

$$\begin{aligned} V(X) &= \sum_O P(O/X)u(O) &\quad O \in \boldsymbol{O} \\ &= \sum_{A,S} P(A \& S/X)u(A \& S) &\quad A \in \boldsymbol{A}, S \in \boldsymbol{S}, \end{aligned}$$

This is X's news value. It measures the average extent to which learning X would provide the agent with evidence for thinking that desirable outcomes will ensue. For propositions that describe past and present events it corresponds to the "amount of happiness" that the agent would derive from reading "X is true" in a reliable newspaper. When X is about the future, one can think of $V(X)$ as the "amount of happiness" she would derive from that learning that an *infallible* soothsayer has predicted that X.

It is important to note that this applies to *actions* as well. When A is a proposition whose truth-value the agent can directly control, $V(A)$ will be high or low depending upon whether she would regard the information that she is going to perform A as a harbinger of desirable or undesirable results. In effect, $V(A)$ measures the extent to which hearing the infallible soothsayer say, "You will do A" would be good or bad news for the agent. Now, since a decision maker, in effect,

---

[4] Where there are uncountably many acts and states this expression will be the integral $V(X) = \int_O u(O)dP_X(O)$ where $\boldsymbol{O}$ is the set of all outcomes.

"makes the news"[5] when she acts, evidential decision theory requires her to *seek the good by seeking good news*. If this seems an odd way to think about actions, let me say that I agree. For the moment, however, let's put this issue to the side and see where Jeffrey's theory leads.

From the formal point of view, the most important fact about Jeffrey's Equation is that it allows us to express the utility of any disjunction as a function of the utilities of its disjuncts.

Theorem 4.1. *If $\{X_1, X_2, X_3, \ldots\}$ is a countable set of pairwise incompatible propositions in Ω and $X = \vee_j X_j$ is their disjunction, then $V(X) = \Sigma_j P(X_j/X)V(X_j)$.*

Proof:
$$
\begin{aligned}
\Sigma_j \, P(X_j/X)V(X_j) &= \Sigma_j P(X_j/X)[\Sigma_{A,\,S} \, P(A \,\&\, S/X_j)u(A \,\&\, S)] \\
&= \Sigma_j \, P(X_j)/P(X)[\Sigma_{A,\,S} \, P(A \,\&\, S/X_j)u(A \,\&\, S)] \\
&= \Sigma_j \, \Sigma_{A,\,S} \, [P(A \,\&\, S \,\&\, X_j)/P(X)]u(A \,\&\, S) \\
&= \Sigma_j \, P(A \,\&\, S/X)u(A \,\&\, S) \\
&= V(X)
\end{aligned}
$$

Here the second equality holds because X is the disjunction of the X_j, and the fourth holds because the X_j are mutually incompatible. ∎

Theorem 4.1 says that the news value of a proposition is always a function of the news values of the *ways in which it might be true*. The important thing to note about the theorem is that its validity does not depend on the particular way in which X happens to be subdivided. If X can be expressed as a disjoint disjunction of both $\{X_1, X_2, X_3, \ldots\}$ and $\{Y_1, Y_2, Y_3, \ldots\}$, then Theorem 4.1 entails that $\Sigma_j P(X_j/X)V(X_j) = \Sigma_k P(Y_k/X)V(Y_k)$. So, while X's news value is always a function of the news values of the various ways in which it might be true, it does not matter how one slices up these ways since the value for $V(X)$ is the same no matter how it is done. One way to put this point is to say that the function V is *partition invariant*.

Partition invariance makes it possible to employ expected utility maximization in small-world decision making. Since the acts, states, and outcomes of a small-world decision problem **D** are disjunctions of the acts, states, and outcomes of the grand-world problem **D**^{G}, it follows from partition invariance that any assignment of probabilities and Jeffrey utilities to the later entities determines a unique assignment of probabilities and Jeffrey utilities to the former. Moreover, in contrast with Savage's theory, an agent who *correctly* estimates her

---

[5] Jeffrey (1965/1983, pp. 83–84).

fully considered desires and beliefs regarding the states and outcomes in **D** can be sure that the policy of maximizing news value in **D** will not lead her astray because, no matter how she chooses to subdivide the possibilities, the use of Jeffrey's Equation will lead her to the same results that she would obtain were she facing the grand-world decision itself. In Jeffrey's theory, then, the basic laws of rationality are the same for both kinds of decision making, and there is guaranteed agreement between grand- and small-world representations of preferences. This guarantee is precisely what Savage could not deliver. The partition invariance of Jeffrey's theory should thus be seen as one of its main advantages over Savage's theory.

4.3 NON-UNIQUENESS

Another point of contrast between Jeffrey's theory and Savage's is that the former leaves a degree of freedom in the choice of a utility representation that is not exhausted when a unit and a zero point are fixed. In Savage's framework, a preference ranking defined over a sufficiently rich family of prospects can be given an expected utility representation (P, u) in which the probability P is unique and the utility u is unique up to a positive linear transformation. This provides an *absolute* scale on which to measure the strengths of the agent's beliefs and an *interval* scale on which to measure the intensities of her desires. This sort of uniqueness cannot be secured within Jeffrey's system. Except in extreme cases, an agent's preferences for news items do not unambiguously determine either her evidential expected utilities or her subjective probabilities. In this section we shall see how this indeterminacy arises.

Start by fixing a zero point and a unit relative to which news value is to be measured. It is natural to set the zero at the necessarily true proposition $\mathbf{T} = (X \vee \neg X)$. Jeffrey's Equation stipulates that $\mathbf{T}$ will have a "news value" given by

$$V(\mathbf{T}) = P(X)V(X) + P(\neg X)V(\neg X)$$

where X may be any proposition in Ω. There is a clear sense, however, in which $\mathbf{T}$ cannot really be news; learning that a tautology is true is not really learning anything, or at least nothing informative. So, it makes sense to set $V(\mathbf{T}) = 0$.

This convention has a number of advantages. First, it amounts to assigning a news value of 0 to any element of Ω whose subjective probability is 1. This reflects the fact that a proposition no longer counts as "news" for someone who is already certain that it is true.

Second, it follows from Theorem 4.1 that $V(X) > V(\neg X)$ implies $V(X) > V(\mathbf{T}) > V(\neg X)$ as long as $1 > P(X) > 0$. This says that an agent who regards X as "good news" will rank X above $\mathbf{T}$ and $\mathbf{T}$ above $\neg X$; "no news" is always worse than "good news" and better than "bad news." Setting $V(\mathbf{T}) = 0$ gives "good news" a positive value and "bad news" a negative value, thereby making it reasonable to think of $\mathbf{T}$ as a description of the current *status quo* and of $V(X)$ as a measure of the desirability of the situation that would be obtained by factoring X's truth into *status quo*.

To fix a unit for measuring utility we need to impose an "anti nirvana" requirement by assuming that there will be at least one proposition whose truth would be good news for the agent and whose falsity would be bad news for her. Specifically, it is required that there be at least one $G \in \Omega$ such that G is strictly preferred to $\mathbf{T}$ and $\mathbf{T}$ is strictly preferred to $\neg G$. (Note that neither G nor $\neg G$ can have a subjective probability of 1 or 0 because, as we have just seen, every probability 1 proposition is ranked with $\mathbf{T}$.) We will conventionally fix some such proposition G and set $V(G) = 1$.

If news value can be measured on an interval scale then our decision to set $V(G) = 1$ and $V(G \vee \neg G) = 0$ should uniquely pick out V. Surprisingly, this will not be the case except in the special circumstance in which V is *unbounded*. To see why, let $(>, \geq)$ be a preference ranking defined over Ω (not just $\mathbf{A}$), and suppose that $(>, \geq)$ is represented by a pair (P, V) in which P is a probability on Ω and V is a conditional utility defined (via Jeffrey's Equation) in terms of P and some underlying assignment u of utilities to the atomic elements of Ω. For any $X, Y \in \Omega$, we then have $X > Y$ only if $V(X) > V(Y)$, and $X \geq Y$ only if $V(X) \geq V(Y)$. Let Λ be the set of all real numbers such that $1 + \lambda V(X) > 0$ for all $X \in \Omega$. Λ is nonempty because it contains $\lambda = 0$. In general, Λ will be the interval defined by

$$-1/\inf\{V(X)\colon X \in \Omega\} \geq \lambda \geq -1/\sup\{V(X)\colon X \in \Omega\}$$

where the endpoints are, respectively, the least upper and greatest lower bounds on the values of V. Λ will thus contain an infinity of values unless V is unbounded both above and below (in which case $\sup\{V(X)\colon X \in \Omega\} = \infty = -\inf\{V(X)\colon X \in \Omega\}$). For each λ between the endpoints consider the functions defined by

Bolker's Equations:

$$P_\lambda(X) = P(X)[1 + \lambda V(X)]$$
$$V_\lambda(X) = V(X)[(1 + \lambda)/(1 + \lambda V(X))]$$

These equations determine a set of pairs (P_λ, V_λ) indexed by elements of Λ. (P, V) is located within this set as (P_0, V_0), and each of the set's elements conforms to our convention to set $V_\lambda(G \vee \neg G) = 0$ and $V_\lambda(G) = 1$.

Ethan Bolker proved the following important result, which shows that each of these pairs is an expected utility representation of $(>, \geq)$.

Theorem 4.2 (Bolker).[6] *For each pair (P_λ, V_λ), with $\lambda \in \Lambda$:*

- P_λ *is a countably additive probability defined on Ω.*
- (P_λ, V_λ) *obeys Jeffrey's Equation.*
- (P_λ, V_λ) *represents $(>, \geq)$ if and only if $(P, V) = (P_0, V_0)$ also represents $(>, \geq)$.*

Proof: I will leave it to the reader to work out for herself why P_λ is a probability.[7] To see why (P_λ, V_λ) obeys Jeffrey's Equation, let us begin by introducing a useful quantity, which Jeffrey calls the *integral* defined by (P_λ, V_λ). It is the product $I_\lambda(X) = P_\lambda(X)V_\lambda(X)$. The following two facts about I are obvious consequences of the definitions given so far:

INT$_1$ Since (P, V) obeys Jeffrey's Equation, I_0 is countably additive, that is,

$$I(\vee_j X_j) = \sum_j I(X_j)$$

for any sequence $\{X_1, X_2, X_3, \ldots\}$ of pairwise incompatible propositions.

INT$_2$ $I_\lambda(X) = I(X)(1 + \lambda)$ for all X in Ω.

---

[6] Kurt Godel obtained this result independently. See Jeffrey (1965/1983, p. 143).
[7] The only nontrivial step is Additivity. Let $\{X_1, X_2, X_3, \ldots\}$ be a set of mutually incompatible propositions whose disjunction is X. Then,

$$P_\lambda(X) = P(X)(1 + \lambda V(X))$$
$$= P(X)\left(1 + \lambda\left[\sum_j P(X_j / X)V(X_j)\right]\right)$$
$$= P(X) + \lambda\left(\sum_j P(X_j)V(X_j)\right)$$
$$= \sum_j P(X_j) + \lambda\left(\sum_j P(X_j)V(X_j)\right)$$
$$= \sum_j P(X_j)(1 + \lambda V(X_j))$$
$$= \sum_j P_\lambda(X_j)$$

124

This second fact makes it easy to show that (P_λ, V_λ) obeys Jeffrey's Equation. For if $\{X_1, X_2, X_3, \ldots\}$ is any sequence pairwise incompatible propositions in Ω, then

$$
\begin{aligned}
V_\lambda(X) &= V(X)[(1 + \lambda)/(1 + \lambda V(X))] && \text{definition of } V_\lambda \\
&= \left[\sum_j P(X_j/X)V(X_j)\right][(1 + \lambda)/(1 + \lambda V(X))] && \text{Theorem 4.1} \\
&= \left[\sum_j I(X_j)/P(X)\right][(1 + \lambda)/(1 + \lambda V(X))] && \text{definition of } I \\
&= \left[\sum_j I(X_j)(1 + \lambda)\right]/[P(X)(1 + \lambda V(X))] && \\
&= \left[\sum_j I(X_j)(1 + \lambda)\right]/P_\lambda(X) && \text{definition of } P_\lambda \\
&= \sum_j I_\lambda(X_j)/P_\lambda(X) && \text{INT}_2 \\
&= \sum_j P_\lambda(X_j/X)V_\lambda(X_j) && \text{definition of } I
\end{aligned}
$$

To show that (P_λ, V_λ) represents $(>, \geq)$ when (P, V) does, one just needs a bit of algebra to establish that all of the following inequalities are equivalent:

$$
\begin{aligned}
&V_\lambda(X) \geq V_\lambda(Y) \\
&V(X)[(1 + \lambda)/(1 + \lambda V(X))] \geq V(Y)[(1 + \lambda)/(1 + \lambda V(Y))] \\
&V(X)(1 + \lambda V(Y)) \geq V(Y)(1 + \lambda V(X)) \\
&V(X) + \lambda V(X)V(Y) \geq V(Y) + \lambda V(X)V(Y) \\
&V(X) \geq V(Y)
\end{aligned}
$$

This completes the proof. ■

   The important thing to notice about Theorem 4.2 is what it does *not* show in the case where V is bounded. Despite agreeing about the preference ordering of propositions in Ω, different (P_λ, V_λ) pairs will not generally agree either about the *ratios of utility differences* among propositions or even about the *comparative* likelihoods of propositions. That is, it is generally not the case that

$$
\frac{V_\lambda(X) - V_\lambda(X^*)}{V_\lambda(Y) - V_\lambda(Y^*)} = \frac{V(X) - V(X^*)}{V(Y) - V(Y^*)}
$$

or that

$$
P_\lambda(X) \geq P_\lambda(Y) \text{ if and only if } P(X) \geq P(Y)
$$

125

The correct relationships are the much more complicated

$$\frac{V_\lambda(X) - V_\lambda(X^*)}{V_\lambda(Y) - V_\lambda(Y^*)} = \frac{V(X) - V(X^*)}{V(Y) - V(Y^*)} \times \frac{(1 + \lambda V(Y))(1 + \lambda V(Y^*))}{(1 + \lambda V(X))(1 + \lambda V(X^*))}$$

and

$$P_\lambda(X) \geq P_\lambda(Y) \text{ if and only if } [P(X) - P(Y)] \geq \lambda[I(Y) - I(X)]$$

It follows from the first of these expressions that we cannot treat an arbitrarily chosen V_λ as an *interval* scale on which the desirabilities of propositions are measured. (Recall from Chapter 1 that *the sine qua non* of interval scales is the invariance of ratios of differences among different scales.) In contrast with Savage's theory, an agent's binary preferences do not uniquely determine the absolute strengths of her desires in Jeffrey's account (unless V happens to be un-bounded).[8] The connection between preferences and beliefs is even looser since the former do not even determine comparative strengths for the latter; that is, different P_λ's do not generate the same likelihood ranking over the propositions in Ω. Evidential decision theory thus allows for the possibility that different sets of beliefs and desires can give rise to the same preference ranking, and this holds even when the set of prospects over which the ranking is defined is very rich.

Even though different (P_λ, V_λ) pairs do not agree about ratios of utility differences or comparative probabilities, they do share one important feature: Ratios of differences of "integrals" are invariant among them. In other words, for any $\lambda \in \Lambda$, one has

$$\frac{I_\lambda(W) - I_\lambda(X)}{I_\lambda(Z) - I_\lambda(Y)} = \frac{I(W) - I(X)}{I(Z) - I(Y)}$$

provided that $I(Z) - I(Y) > 0$. This tells us that, while an agent's preferences for news cannot "reveal" either her basic news values or her subjective probabilities, they can reveal the *product* of the two. Within Jeffrey's theory P and V are "hidden variables" that are only

---

[8] It should be noted in this context that the unboundedness of a preference ranking is not merely a matter of the way in which it happens to be represented. Jeffrey shows how to formulate a qualitative constraint on (>, ≥) that will be satisfied if and only if *every* (P, V) pair that represents it contains a utility that is unbounded both from above and from below. See Jeffrey (1965/1983, p. 142) for details.

reflected in preference through their cumulative effect, which is measured by I.

We shall return to the issue of non-uniqueness, after we have seen how Ethan Bolker was able to prove a representation theorem for evidential decision theory.

4.4 BOLKER'S REPRESENTATION THEOREM

In the mid-1960s Ethan Bolker proved a general mathematical result, relating to functions that can be expressed as quotients of measures, that provides a representation theorem for Jeffrey's decision theory.[9] Bolker, who was concerned more with mathematics than with decision making, did not distinguish among acts, states, and outcomes. He merely assumed an algebra of propositions Ω over which probabilities and utilities would be defined, posited an ordering $(>, \geq)$ over elements of Ω, and inquired after the conditions under which there would exist a probability P defined on Ω and a utility u defined on atomic elements W of Ω such that $V(X) = \Sigma_W P(W/X)u(W)$ would represent $(>, \geq)$. The conditions he discovered did not, of course, secure the existence of a *unique* representation, but they did isolate a *single* class of (P_λ, V_λ) pairs, equivalently (P_λ, u_λ) pairs, that are related by Bolker's Equations. Bolker's Theorem thus provides a representation for $(>, \geq)$ that is as close to unique as it can be given the nature of the representing function.

Like Savage, Bolker assumed that $(>, \geq)$ would completely order Ω. He was thus led to work with $\geq$ as the only primitive term of his theory, and he sought a representation in which $V(X) \geq V(Y)$ would be necessary and sufficient for $X \geq Y$. Once again, since I do not want to assume that the preference rankings of rational agents need to be complete, I will formulate the theory in terms of both $>$ and $\geq$, so as to leave open the "no preference" option of neither strictly preferring X to Y nor weakly preferring Y to X.[10]

The first three axioms are, for all practical purposes, identical to those used in Savage's Theorem:

EDT$_1$ (Nontriviality): $G > (G \lor \neg G) > \neg G$ for some $G \in \Omega$.
EDT$_2$ (Partial Ordering): The decision maker's preferences *partially order* the propositions in Ω; that is for any $X, Y, Z \in \Omega$ one has

[9] Bolker (1966).
[10] The set of axioms I use in Bolker's Theorem are taken, with minor modifications from Jeffrey (1978). The names have been chosen to relate each to its closest cousin in Savage's system. Again, structure axioms are preceded by a "*."

- *Reflexivity of Weak Preference*: $X \geq X$
- *Irreflexivity of Strict Preference*: Not $X > X$
- *Consistency*: $X > Y$ only if $X \geq Y$
- *Transitivity*: If $X \geq Y$ and $Y \geq Z$ then $X \geq Z$
  If $X > Y$ and $Y \geq Z$ then $X > Z$
  If $X \geq Y$ and $Y > Z$ then $X > Z$

*EDT$_3$ (Completeness): The decision maker has a definite preference with regard to every pair of propositions in Ω, so that either $X > Y$ or $X \geq Y$ holds for every $X, Y \in \Omega$.

The first of these says that some proposition is preferred to its negation. If this condition is violated (in the presence of the other axioms), then the decision maker will be indifferent between every prospect and the status quo, so that $X \approx (X \vee \neg X)$ for all $X \in \Omega$. Nothing she could possibly discover, about herself or the world, would matter to her happiness – she would have achieved nirvana. Decision theory is silent about such creatures. The second two axioms together guarantee that the agent's preferences totally order Ω. Again, I have used two conditions here where Bolker has only one because I want to be careful to maintain a clear distinction between structure axioms like *EDT$_3$ and axioms of pure rationality like EDT$_2$.

The next two axioms have no direct analogue in Savage's system. Together they replace his Independence and Stochastic Dominance principles. Their content is not the same, however, because both axioms incorporate the idea that actions can affect the probabilities of states.

EDT$_4$ (Averaging): If X and Y are mutually incompatible propositions such that $X \geq Y$, then $X \geq (X \vee Y) \geq Y$.

EDT$_5$ (Impartiality): If $X \geq X^*$ and if Y and Y^* are each incompatible with X and with X^*, then the following pattern of preferences never occurs

$$Y > (X^* \vee Y) \geq (X \vee Y) > X \geq X^* > (X^* \vee Y^*) \geq (X \vee Y^*) > Y^*$$

unless each of the weak preferences $\geq$ is really an indifference.

Averaging is one way of saying that an agent's desire for the truth of propositions has to be a function of the ways in which they can be true. If one thinks of X and Y as the possible "outcomes" of $X \vee Y$, then the axiom says that a prospect cannot be more desirable than its most desirable outcome, or less desirable than its least desirable outcome.

This makes perfect sense given that we are trying to capture the concept of news value. If an agent weakly prefers X to Y, then she should be at least as happy to learn $X \vee Y$ as she is to learn Y because $X \vee Y$ leaves the possibility of X open. By the same token, she should be happier to learn X than to learn $X \vee Y$ because the latter leaves the possibility of Y open.

Impartiality is the key to the whole theory. The version I have stated here is slightly more general than the one Bolker uses. The two formulations are equivalent in the presence of the other axioms, but it will be useful to have the more general principle around when we come to ask whether the other axioms can be weakened. To understand what EDT_5 says one needs to see why the pattern of preferences it describes would not be compatible with any evidential expected utility representation. Suppose that it was, so that there would be a (P, V) pair obeying Jeffrey's Equation such that

$$(\#) \quad V(Y) > V(X^* \vee Y) \geq V(X \vee Y) > V(X) \geq V(X^*)$$
$$> V(X^* \vee Y^*) \geq V(X \vee Y^*) > V(Y^*)$$

Writing out the leftmost weak inequality using Jeffrey's Equation we get that

$$P(X/X^* \vee Y)V(X^*) + P(Y/X^* \vee Y)V(Y)$$
$$\geq P(X/X \vee Y)V(X) + P(Y/X \vee Y)V(Y)$$

Since conditional probabilities are additive this entails

$$P(X^*/X^* \vee Y)[V(X^*) - V(Y)] \geq P(X/X \vee Y)[V(X) - V(Y)]$$

But (#) tells us that $[V(X^*) - V(Y)] \leq [V(X) - V(Y)] < 0$, so this can be true only if

$$P(X^*/X^* \vee Y) \leq P(X/X \vee Y)$$

(or, more precisely, if $P(X^*/X^* \vee Y)/P(X/X \vee Y) \leq [V(X) - V(Y)]/[V(X^*) - V(Y)] \leq 1$). A simple calculation then shows that this last inequality will hold if and only if $P(X)$ is at least as large as $P(X^*)$. So, $P(X) \geq P(X^*)$ must hold because $V(X^* \vee Y) \geq V(X \vee Y)$. An identical line of reasoning shows that $P(X^*) \geq P(X)$ must hold because $V(X^* \vee Y^*) \geq V(X \vee Y^*)$. What (#) describes, then, is a situation in which both $P(X) \geq P(X^*)$ and $P(X^*) \geq P(X)$ hold. This is fine if $P(X) = P(X^*)$. However, all of the preceding inferences go through

129

with weak inequalities replaced by strict ones when any of the inequalities in (#) is strict. Thus, if Jeffrey's Equation is to assign consistent probabilities to propositions in Ω, then the preference pattern described in the Impartiality Axiom must be prohibited unless each $\geq$ is replaced by an $\approx$. Otherwise the agent is committed to being more confident in X than in X^* and to being as confident in X^* as in X.

To see why Impartiality is a reasonable constraint on preferences for news imagine a horse race in which the entries are Secretariat, Trigger, and Stewball. You have placed bets in such a way that you stand to gain $10 if Secretariat wins ($Y$), $5 if Trigger wins ($X$), and nothing if Stewball wins (X^*). Suddenly you hear that one of the latter two horses has thrown a shoe and cannot run. Under what conditions would you prefer it to be Trigger who threw it? Answer: If you think that Trigger is much more likely than Stewball to beat Secretariat, then you might prefer to have Stewball in the race (despite the risk of losing $5) because this makes it more likely that Secretariat will win. But, if you feel that Stewball has a better chance of beating Secretariat than Trigger does, then you will definitely not want Stewball to run. The basic point is that the preferences $Y > (X^* \vee Y) \geq (X \vee Y) > X > X^*$ make sense only if you are more confident in X's truth than in X^*'s. Similarly, $X > X^* > (X^* \vee Y^*) \geq (X \vee Y^*) > Y^*$ makes sense only if are more confident in X^* than in X. Thus, you should only hold both these patterns of preferences simultaneously if you are both more confident in X than in Y and more in X^* than in X, which you cannot be.

Impartiality allows us to characterize the relationship of coherence that must hold between a preference ranking and a comparative likelihood ranking for the two to be jointly represented by a (P, V) pair obeying Jeffrey's Equation. In this regard EDT$_5$ is like Savage's Independence Axiom. Here is the appropriate coherence condition:

Coherence. If X, X^*, $Y \in \Omega$ and Y is incompatible with both X and X^*, then

- $X .\geq. X^*$ holds when either $Y > (X^* \vee Y) \geq (X \vee Y) > X \geq X^*$ or $X^* \geq X > (X^* \vee Y) \geq (X \vee Y) > Y$.
- $X .=. X^*$ holds when $Y > (X^* \vee Y) \approx (X \vee Y) > X \approx X^*$ or $X^* \approx X > (X^* \vee Y) \approx (X \vee Y) > Y$.
- $X .>. X^*$ holds when either $Y > (X^* \vee Y) \geq (X \vee Y) > X \geq X^*$ or $X^* \geq X > (X^* \vee Y) \geq (X \vee Y) > Y$ provided that at least one $\geq$ is replaced by $>$.

In light of this, Impartiality can be read as a principle of consistency for comparative beliefs. It says that an agent's preferences should

130

never commit her to both $X \mathrel{.\geq.} X^*$ and $X \mathrel{.>.} X^*$. When $(.>., .\geq.)$ satisfies Coherence I will say that it *coheres* with $(>, \geq)$.

As in the case of Savage's version of this principle, I want to caution the reader against regarding Coherence as a *definition* of conditional likelihood. The same reasons given there apply here; reading the principle as a definition smacks of behaviorism, and it engenders a brand of pragmatism that refuses to recognize that rational belief is governed by *epistemic* laws having to do with the pursuit of truth, not with pragmatic laws having to do with the pursuit of happiness. There is an additional reason for caution here, however. In Savage's theory, with its highly structured set of "acts," it was possible for an agent's preferences to determine her beliefs *completely*. As we saw in the last section, that is not true here. A complete preference ranking over Ω will typically be consistent with many different likelihood rankings. This is why the version of Coherence just stated was not expressed in terms of biconditionals; in evidential decision theory there simply is no way to fill in the blank in "$X \mathrel{.>.} X^*$ only if $(>, \geq)$ is such that ——" because *no* patterns of the types described in Coherence need to hold in a rational preference ranking. Given this, it follows that one can read Coherence as a definition only by taking it to be a "partial" specification of meaning of the sort that Carnap has discussed in various places.[11] One then has two choices about how to view cases in which $(>, \geq)$ does not dictate either $X \mathrel{.>.} X^*$ or $X \mathrel{.\geq.} X^*$. One can either (i) suppose that there are further facts about the agent, which go beyond her preferences, that determine whether she is more confident in X or in X^*, or (ii) deny that the agent has a definite view about the relative likelihoods of X and X^* *merely* because no such opinion is reflected in her preferences. The former maneuver grants that pragmatism is false. The latter is a desperate attempt to save it. It says, in effect, that a person can only be more confident in X than in X^* if there is some third proposition Y, incompatible with both, that she desires sufficiently strongly to make one of the preference patterns in Coherence true. This is clearly absurd; the question of whether to believe X or X^* more strongly is a matter of the evidence that one has for either proposition, not of the strength of one's desires for things that can only come about when both are false. Thus, Coherence should be read not as a *definition* of conditional likelihood, but as an expression of a relation between conditional likelihood and rational preference.

[11] See, for example, Carnap (1936).

We now need an analogue of Savage's nullity principle. Here is the appropriate definition of a null proposition in Jeffrey's system:

Definition. *X is* null *relative to* $(>, \geq)$ *if and only if* $(X \vee Y) \approx Y$ *for some $Y \in \Omega$ that is incompatible with X and for which either $X > Y$ or $Y > X$ holds.*

As before, denote the set of null propositions relative to $(>, \geq)$ as ***N***. It should be clear that a rational agent who is not indifferent between X and Y will only be indifferent between receiving news of $X \vee Y$ and receiving news of Y if she is certain that X is false. Likewise, when she is sure that X is false, news of $X \vee Y$ will just *be* news of Y from her perspective, so she will be indifferent between the two prospects. This makes the following requirement reasonable:

EDT$_6$ (Nullity): If X is null, then $(X \vee Y) \approx Y$ for all $Y \in \Omega$.

It is a consequence of this that a decision maker is always indifferent between prospects that differ by only a null proposition, so that $(X \mathbin{\&} \neg Y) \vee (\neg X \mathbin{\&} Y) \in$ ***N*** implies that $X \approx Y$.

We need the following three requirements to guarantee that the decision maker's beliefs will be representable by a countably additive probability function.

EDT$_7$ (Null Consistency): The set of all null statements in Ω is a σ-ideal:

- $(X \mathbin{\&} \neg X) \in$ ***N***
- If $X \in$ ***N*** and $Y \in \Omega$, then $X \mathbin{\&} Y \in$ ***N***
- If $\{X_1, X_2, X_3, \ldots\}$ is a countable subset of ***N***, then $X = \vee_j X_j$ is also in ***N***

EDT$_8$ (Archimedean Axiom): Any collection of pairwise incompatible propositions in $\Omega \sim$ ***N*** is countable.

EDT$_9$ (*Continuity*): Let $\{X_1, X_2, X_3, \ldots\}$ be a countable set of pairwise incompatible propositions in Ω whose disjunction is X. Then,

$$\text{If } Y \geq (X_1 \vee X_2 \vee \ldots \vee X_n) \text{ for all } n \text{ then } Y \geq X$$
$$\text{If } (X_1 \vee X_2 \vee \ldots \vee X_n) \geq Y \text{ for all } n \text{ then } X \geq Y$$

EDT$_7$, which ends up being a derived principle in Savage's theory, says that elements of ***N*** really do behave like 0 probability events. EDT$_8$ rules out infinitesimal probabilities. EDT$_9$ is a generalization of Averaging. It ensures that no proposition can be strictly more (less) desirable than the most (least) desirable way in which it might be true. Among other things, this guarantees that the probability in each representing (P, V) pair will be countably additive.

Like Savage before him, Bolker needed to assume that any event could be subdivided into subevents of arbitrarily small probability.

*EDT$_9$ (*Richness*): $\Omega \sim \boldsymbol{N}$ is nonatomic. That is, if X is nonnull, then there exist incompatible nonnull propositions X_1 and X_2 such that $X = (X_1 \vee X_2)$.

This is a structure axiom. It is not strictly necessary for the existence of the desired representation, but it is essential to Bolker's construction of such a representation. As with any structure condition, it should be read as a requirement of coherent extendibility.

On the basis of these axioms Bolker was able to establish:

Theorem 4.3a (Bolker's Theorem, Existence). *Let a decision maker's preferences among news items in a σ-complete Boolean algebra Ω be described by the ranking (>, $\geq$). If (>, $\geq$) satisfies the Jeffrey/Bolker axioms EDT$_1$–*EDT$_9$, then there exists at least one countably additive probability P defined on Ω and one utility u defined on the atomic elements W of Ω whose associated conditional expected utility function $V(X) = \Sigma_W P(W/X)u(W)$ both obeys the scaling convention that $V(G) = 1$ and $V(G \vee \neg G) = 0$, and represents the agent's desires for news in the sense that $X > X^*$ only if $V(X) > V(X^*)$ and $X \geq X^*$ only if $V(X) \geq V(X^*)$.*[12]

Theorem 4.3b (Bolker's Theorem, Semi-Uniqueness). *If (>, $\geq$) satisfies the Jeffrey/Bolker axioms and if (P_0, V_0) and (P, V) are probability/ utility pairs that represent (>, $\geq$) and that obey the scaling convention, then there is a real number λ that falls between $-1/\inf\{V(X): X \in \Omega\}$ and $-1/\sup\{V(X): X \in \Omega\}$ such that*

- $P(X) = P_\lambda(X) = P_0(X)[1 + \lambda V_0(X)]$
- $V(X) = V_\lambda(X) = V_0(X)[(1 + \lambda)/(1 + \lambda V_0(X))]$

for all $X \in \Omega$.

The second part of the theorem shows that the class of (P, V) pairs that represent (>, $\geq$) is as small as it can be given the indeterminacy inherent in Jeffrey's Equation. It says that every representation for the decision maker's preference ranking will be found among the (P_λ, V_λ) pairs associated with any fixed representation (P_0, V_0).

---

[12] These conditionals can be replaced by biconditionals given that the preference ranking has been assumed to be complete.

One must be careful not to read too much into this "uniqueness." It is important to understand how weak a constraint it imposes on a decision maker's beliefs. There is a tendency to think that the Jeffrey/Bolker axioms, like those of Savage, can only be satisfied by someone whose judgments of relative likelihood can be probabilistically represented. Savage's axioms, of course, were sufficient to ensure the existence of a unique probability representation. While this cannot occur here, one tends to read the Jeffrey/Bolker axioms as ensuring that *at least one* such representation will exist among the P_λ, that is, that there will be at least one λ such that the $P_\lambda(X) \geq P_\lambda(X^*)$ holds whenever the decision maker is as confident in X as she is in X^*. This is not so. It is possible to have preferences that satisfy the Jeffrey/Bolker axioms even if one's beliefs are *probabilistically incoherent*. This is not an obvious point, and it will be worth our while to spend some time seeing why it is true.

The first thing to understand is that the probabilities in (P_λ, V_λ) pairs that represent $(>, \geq)$ will not agree about the comparative likelihoods of all the propositions in Ω (unless V is unbounded). This is a consequence of the following fact about probability representations:

Theorem 4.4 (Ordinal Uniqueness Theorem). *Let P and P* be probability functions defined on Ω and suppose that*

- *P is atomless in the sense that $P(X) > 0$ only if there are mutually incompatible propositions X_1 and X_2 in Ω such that $X = (X_1 \vee X_2)$ where $P(X_1)$ and $P(X_2)$ are nonzero.*
- *P and P* are ordinally similar in the sense that $P(X) \geq P(X^*)$ iff $P^*(X) \geq P^*(X^*)$ for all $X, X^* \in \Omega$.*

Then, P and P* are identical.

Proof: The proof is based on the following general fact, which was established by Savage:[13]

Lemma. *If P is atomless, then for any $X \in \Omega$ and any real number $0 \leq c \leq 1$ there is a proposition X_C that entails X such that $P(X_C) = cP(X)$.*

To prove Theorem 4.4, begin by applying this lemma to the necessary proposition $X = \mathbf{T}$ with $c = 1/n$ for some integer $n > 0$. This yields a proposition $X(1, n) \in \Omega$ such that $P(X(1, n)) = 1/n$. Next apply the lemma to $\neg X(1, n)$ with $c = 1/n$ to get a proposition $X(2, n)$ mutually

[13] See Savage (1954/1972, p. 37).

incompatible with $X(1, n)$ such that $P(X(2, n)) = 1/n$. Continuing this way, we get a partition $\{X(1, n), X(2, n), \ldots, X(n, n)\}$ such that $P(X(j, n)) = 1/n$ for $j = 1, 2, \ldots, n$. Assuming that P* is ordinally similar to P, and because the $X(j, n)$ form a partition, it follows that $P^*(X(j, n)) = 1/n$ for all j.

Now, for $m \leq n$, define $Z(m, n) = (X(1, n) \vee X(2, n) \vee \ldots \vee X(m, n))$, and note that $P(Z(m, n)) = P^*(Z(m, n)) = n/m$. Since we can carry out this construction for every value of n, we are left with a set of propositions $\{Z(r): r$ a *rational* number in $[0, 1]\}$ such that $P(Z(r)) = P^*(Z(r)) = r$. Since the rationals are dense in $[0, 1]$ it follows that for any $X \in \Omega$ we can find two sequences of rationals $\{r_j{}^+\}$ and $\{r_j{}^-\}$ that converge to $P(X)$ monotonically from above and below. From this it follows that

$$P\left(Z\left(r_1{}^+\right)\right) \geq P\left(Z\left(r_2{}^+\right)\right) \geq P\left(Z\left(r_3{}^+\right)\right) \geq \ldots \geq P(X) \geq \ldots$$
$$\geq P\left(Z\left(r_3{}^-\right)\right) \geq P\left(Z\left(r_2{}^-\right)\right) \geq P\left(Z\left(r_1{}^-\right)\right)$$

and by ordinal similarity that

$$P^*\left(Z\left(r_1{}^+\right)\right) \geq P^*\left(Z\left(r_2{}^+\right)\right) \geq \ldots \geq P^*(X) \geq \ldots$$
$$\geq P^*\left(Z\left(r_2{}^-\right)\right) \geq P^*\left(Z\left(r_1{}^-\right)\right)$$

Consequently, since $P^*(Z(r)) = P^*(Z(r)) = r$, and since $\{r_j{}^+\}$ and $\{r_j{}^-\}$ converge to $P(X)$ monotonically from above and below it follows that $P(X) = P^*(X)$ as desired. This completes the proof. ∎

In the present context, the important thing about the Ordinal Uniqueness Theorem is that it prevents different (P_λ, V_λ) pairs from agreeing about the comparative likelihoods of propositions in Ω. Suppose that $(>, \geq)$ obeys the Jeffrey/Bolker axioms and is represented by (P, V). The richness condition *EDT_9 ensures that (P, V), and every one of the (P_λ, V_λ) pairs associated with it, is nonatomic. For any two such pairs (P_λ, V_λ) and (P_μ, V_μ), Ordinal Uniqueness requires the existence of propositions such that $P_\lambda(X) > P_\lambda(X^*)$ and $P_\mu(X) \leq P_\mu(X^*)$. This means that, except in the special case in which V is unbounded, an infinite number of comparative likelihood rankings, one for each P_λ, will be compatible with any preference ranking that obeys the Jeffrey/Bolker axioms. The kind of compatibility at issue here is the kind expressed in Coherence. Given what has already been said, it should be obvious that all the likelihood rankings defined by

135

$$X .\geq_{\cdot\lambda} X^* \text{ if and only if } P_\lambda(X) \geq P_\lambda(X^*)$$

will cohere with (>, ≥).

Even more interesting is the fact that there are likelihood rankings that cohere with (>, ≥) that cannot be represented by *any* probability. To see why this is so suppose (P_λ, V_λ) and (P_μ, V_μ), $\lambda \neq \mu$, both represent (>, ≥). Define a likelihood ranking (.>., .≥.) by setting

$$X .\geq. X^* \text{ when } X .\geq_{\cdot\lambda} X^* \text{ and } X .\geq_{\cdot\mu} X^*$$

$$X .>. X^* \text{ when } X .>_{\cdot\lambda} X^* \text{ and } X .>_{\cdot\mu} X^*$$

$$X .>. X^* \text{ otherwise}$$

Since there will be propositions X and X^* such that $X .\geq_{\cdot\lambda} X^*$ and $X^* .>_{\cdot\mu} X$ this new ranking will contain both $X .>. X^*$ and $X^* .>. X$, which prevents it from being represented by any probability. However, this fact will never show up in the agent's preferences for news. The ranking (.>., .≥.) satisfies Coherence because it agrees with the coherent rankings $(.>_{\cdot\lambda}, .\geq_{\cdot\lambda})$ and $(.>_{\cdot\mu}, .\geq_{\cdot\mu})$ in any case where the principle can be applied. The general point, then, is that the Jeffrey/Bolker axioms do not constrain a decision maker's beliefs enough to uniquely determine her estimates of comparative likelihood or even to force these estimates to be probabilistically representable.

Many decision theorists will see this as a serious flaw in Jeffrey's theory. The orthodox view, after all, is that a specification of an agent's binary preferences should always suffice to determine the strengths of her desires and beliefs completely. This idea seems appealing because it suggests the possibility of a theory of epistemic rationality that fully characterizes the laws of rational belief merely by codifying the laws of rational preference. What we have here is yet another instance of the kind of pragmatism that I argued against in earlier chapters, and it is just as untenable here as it was there. In fact, when one thinks about the matter carefully, it is clear Jeffrey's theory has it exactly right. Any version of expected utility theory is committed to two theses: (i) the expected utility of any proposition is determined by the utilities and probabilities of the various relevant ways in which it could be true, and (ii) the effect of these utilities and probabilities on the utility of the whole is a function of the magnitude of their *product*. Utility theorists differ about which "ways of being true" are "relevant" and about what probabilities should be used to compute expected values, but (i) and (ii) are not up for grabs; to deny them is to cease advocating expected utility maximization. Given this,

136

one should expect that the *only* thing that will be directly determined by an agent's preferences will be quantities like the "interval" that are ultimately analyzable in terms of utility/probability products. While it is true that one can sometimes factor out the effects of P and u given a rich enough set of preferences to work with, this should not be taken to suggest that the possibility of so factoring is somehow a requirement of rationality. Thinking that it is leads theorists to impose absurdly strong structural requirements on preference rankings, like the existence of "constant" acts, that purchase the possibility of inferring beliefs and desires from preferences at the expense of making the theory entirely unrealistic.

On the other hand, while Jeffrey's theory does provide a correct analysis of preference for news as far as it goes, it is incomplete in an important way. The fact that the Jeffrey/Bolker axioms do not impose a sufficiently strong constraint on a decision maker's beliefs to ensure that they can be probabilistically represented clearly indicates that something has been left out. One can, of course, supply this extra "something" by requiring rational preferences to be unbounded. But this seems ad hoc; the boundedness or unboundedness of preferences is simply not relevant to the question of whether one's beliefs should conform to the laws of probability. Moreover, as we saw in our discussion of Pascal's wager in Chapter 1, unbounded utility functions generate seemingly insoluble problems, such as the St. Petersburg Paradox. Indeed, it may not be rational to have preferences that can only be represented by unbounded utility functions. The bottom line is that Jeffrey/Bolker axioms need to be augmented, not with further constraints on rational preference, but with further constraints on rational belief. Jeffrey himself seems to agree:

This underdetermination of V and P by the preference relation . . . is fascinating, but may be seen as a flaw. (Must one have preferences of unlimited intensities in order to have a perfectly sharp subjective probability measure?) If so, the flaw is removed by using two primitives: preference and comparative probability. With these primitives, one ought to be able to drop some of Bolker's restrictions on preference in favor of conditions on comparative probability and conditions connecting preference and comparative probability. I would expect that in this way one could get significantly closer to an existence theorem in which the conditions are necessary as well as sufficient for existence of V and P, while obtaining the usual uniqueness result: P is unique, and V is determined up to a positive affine transformation. It would be a job worth doing.[14]

---

[14] Jeffrey (1974, pp. 77–78).

In this section I will carry out the program Jeffrey suggested in the preceding excerpt by finding systems of constraints on preferences and comparative beliefs that together suffice to ensure a unique joint probability/utility representation.[15] Let's start out by requiring the decision maker's estimates of comparative likelihood to satisfy the axioms of comparative probability from the previous chapter.

CP_1 (*Normalization*): $(X \vee \neg X) \mathrel{.>.} (X \mathbin{\&} \neg X)$

CP_2 (*Boundedness*): $(X \vee \neg X) \mathrel{.\geq.} X \mathrel{.\geq.} (X \mathbin{\&} \neg X)$

CP_3 (*Ranking*): $(.>., .\geq.)$ is a partial ordering of Ω

*CP_4 (*Completeness*): $X \mathrel{.>.} Y$ or $Y \mathrel{.\geq.} X$ for any $X, Y \in \Omega$

CP_5 (*Quasi additivity*): If Y is incompatible with both X and X^*, then $X \mathrel{.>.} X^*$ if and only if $(X \vee Y) \mathrel{.>.} (X^* \vee Y)$, and $X \mathrel{.\geq.} X^*$ if and only if $(X \vee Y) \mathrel{.\geq.} (X^* \vee Y)$

CP_6 (*Archimedean Axiom*): If $\mathbf{C}$ is a partition in Ω such that $X \mathrel{.=.} X^*$ for every $X, X^* \in \mathbf{C}$, then $\mathbf{C}$ is finite.

CP_7 (*Continuity*): If $\{X_1, X_2, X_3, \ldots\}$ is a countable set of mutually incompatible propositions in Ω and if $X^* \mathrel{.\geq.} (X_1 \vee \ldots \vee X_n)$ for all n, then $X^* \mathrel{.\geq.} \vee_j X_j$.

As was mentioned in the last chapter, the satisfaction of these axioms plus the condition

*CP_8 (*Non-Atomicity*): If $X \mathrel{.>.} (X \mathbin{\&} \neg X)$, then there are mutually incompatible $X_1, X_2 * \in \Omega$ such that $X = (X_1 \vee X_2)$ and $X_1 \mathrel{.>.} (X \mathbin{\&} \neg X)$ and $X_2 \mathrel{.>.} (X \mathbin{\&} \neg X)$.

---

[15] Just after this manuscript was submitted for publication I discovered that Richard Bradley (1998) has solved the non-uniqueness problem in a way rather different from the one pursued here. Instead of imposing constraints on comparative beliefs as I do, Bradley introduces a special conditional operator "$\rightarrow$" and assumes that rational agents typically have preferences about the truth of conditionals defined using this operator. The formal result he obtains is quite interesting, but I must admit to some skepticism about the existence of the required conditionals, which have some rather remarkable properties. For the result to go through, the unconditional probability of a Bradley conditional is the probability of its consequent conditional on its antecedent, so that $P(X \rightarrow Y) = P(Y/X)$. Moreover, its news value must be $V(X \rightarrow Y) = V(X \supset Y) + V(X \mathbin{\&} \neg Y)P(X \mathbin{\&} Y)$. As we shall see in Chapter 6, there are serious obstacles to the first of these identities. More importantly, it is not clear to me how to interpret the second summand in the expression for $V(X \rightarrow Y)$. On the face of things, it seems that a quantity like $V(X \mathbin{\&} \neg Y)P(X \mathbin{\&} Y)$, which weights the news value of a prospect with the probability of a *logically incompatible* prospect, should have nothing whatever to do with what a person wants or does. Nevertheless, I do not want to judge Bradley's work prematurely since my reactions to it are not yet fully thought out.

ensures that there will be a *unique* countably additive probability that represents (.>., .≥.). This result will hereafter be referred to as

Villegas's Theorem. *If (.>., .≥.) satisfies CP_1–$*CP_8$, there exists a unique countable additive probability function P* defined on Ω such that P*(X) ≥ P*(Y) if and only if X .≥. Y.*

We shall call a preference ranking (>, ≥) and a comparative likelihood ranking (.>., .≥.) *EDT-coherent* exactly if (>, ≥) obeys the Jeffrey/Bolker axioms EDT_1–$*EDT_9$, (.>., .≥.) satisfies the axioms of comparative probability CP_1–$*CP_8$, and (.>., .≥.) *coheres* with (>, ≥) in the sense that the rankings jointly obey Coherence. There are a number of redundancies in these requirements, but for the moment let's use them as is to see where they lead. The theorem I will prove is this:

Theorem 4.5. *If (.>., .≥.) and (>, ≥) are EDT-coherent rankings defined on some σ-algebra Ω, then there is a unique pair (P, u) consisting of a countably additive probability P on Ω and a real-valued utility u defined on atomic elements W of Ω such that*

- *P represents (.>., .≥.)*
- *The conditional expected utility function $V(X) = \Sigma_W P(W/X)u(W)$ both represents (>, ≥) and obeys the scaling convention that V(G) = 1 and V(G ∨ ¬G) = 0.*

It might seem as if this theorem does not stand in need of proof at all. After all, since (>, ≥) satisfies the Jeffrey/Bolker axioms, Bolker's Theorem supplies us with a set of (P_λ, V_λ) pairs whose second term represents this ranking. Similarly, by virtue of CP_1–CP_8 we can use Villegas's theorem to obtain a unique probability representation P* of (.>., .≥.). Why, one might wonder, isn't this enough to do the job? The answer it that, for all we have said so far, there is no reason to think that P* corresponds to any of the P_λ, which means that there is no reason to think that (>, ≥) and (.>., .≥.) have a *joint* representation. To show that such a joint representation exists we need to find a constant λ between $-1/\inf\{V(X): X \in \Omega\}$ and $-1/\sup\{V(X): X \in \Omega\}$ such that $P* = P_\lambda$. This turns out to be surprisingly difficult. It is the burden of the next five lemmas.

The first three are due to Bolker and will be stated without proof. Two of them involve the notion of an *additive set*.[16] A subset Γ of Ω is *additive* exactly if it is closed under finite proper differences, so that

---

[16] Bolker (1966, p. 292).

$(X \mathbin{\&} \neg Y) \in \Gamma$ when $(X \vee Y)$, $Y \in \Gamma$, and under arbitrary disjoint disjunctions, so that $\vee_\alpha X_\alpha \in \Gamma$ whenever $\{X_\alpha\}$ is a family of pairwise incompatible propositions in Γ. Γ is *atomless* if each $X \in \Gamma$ can be partitioned into mutually incompatible propositions X_1, $X_2 \in \Gamma$. The crucial facts about atomless additive sets are:[17]

Lemma 4.5a. *Suppose that Γ is an atomless additive set in Ω, and let P and P^* be probabilities on Ω such that $P(X)$ and $P^*(X)$ are nonzero for all $X \in \Gamma$. Suppose further that $P(X) = P(Y)$ if and only if $P^*(X) = P^*(Y)$. Then, there exists a positive real number c such that $P = cP^*$.*

Lemma 4.5b. *Suppose $(\mathbf{>}, \mathbf{\geq})$ satisfies the Jeffrey/Bolker axioms. Then, for each $Z \in \Omega \sim \mathbf{N}$, the associated $\approx$-equivalence class $\Gamma_Z = \{X \in \Omega \sim \mathbf{N}\!: X \approx Z\}$ is an atomless additive set.*

There are two consequences of these results that are important to what follows. First, since $(.{>}., .{\geq}.)$ is coherent with $(\mathbf{>}, \mathbf{\geq})$ it follows that the null-ideal $\mathbf{N}$ can be characterized either as $\{X \in \Omega\!: X \approx (X \vee Y)\}$ or as $\{X \in \Omega\!: X .=. (X \mathbin{\&} \neg X)\}$. By Lemma 4.5b, this means that any $\approx$-equivalence class Γ_Z will be atomless with respect to *both* $(.{>}., .{\geq}.)$ and $(\mathbf{>}, \mathbf{\geq})$, so that for every $X \in \Gamma_Z$ there will be mutually incompatible propositions X_1, $X_2 \in \Gamma_Z$ such that

- $X_1 \mathbin{\mathbf{<}\,\mathbf{>}} (X_1 \vee Y_1)$ for some Y_1
- $X_2 \mathbin{\mathbf{<}\,\mathbf{>}} (X_2 \vee Y_2)$ for some Y_2
- $X_1 .{>}. (X_1 \mathbin{\&} \neg X_1)$
- $X_2 .{>}. (X_2 \mathbin{\&} \neg X_2)$.

where $\mathbf{<}\,\mathbf{>}$ means "is strictly preferred to or strictly dispreferred to." The second point is that since Γ_Z is atomless with respect to $(.{>}., .{\geq}.)$ it will always be possible to partition any X in Γ_Z into any number of equiprobable parts. We will make this its own principle.

Corollary 4.5c ("Splitting"). *Given any $X \in \Gamma_Z$ and any $n = 1$, 2, 3, ..., one can partition X into n pairwise incompatible propositions $X_1, X_2, \ldots, X_n \in \Gamma_Z$ such that $X_1 .=. X_2 .=. \ldots .=. X_n .{>}. (X \mathbin{\&} \neg X)$.*[18]

In addition to the two lemmas given, Bolker proved the following result, which hinges crucially on the non-atomicity of $\Omega \sim \mathbf{N}$.

---

[17] The proofs can be found in Bolker (1966). The first is his Lemma 2.6, and the second follows from his Theorem 1.11 and Corollary 1.15.

[18] Of course, each X_j will be ranked with X since it will be in Γ_Z.

Lemma 4.5d (The "Solvability" Lemma).[19] *Suppose that X_1 and X_2 are nonnull propositions in Ω, and imagine further that $X_1 > Y > X_2$. Then, there are nonnull propositions $X_1{}^*$ and $X_2{}^*$, which entail X_1 and X_2, respectively, such that $X_1 \approx X_1{}^* > (X_1{}^* \vee X_2{}^*) \approx Y > X_2 \approx X_2{}^*$.*

Quantitatively speaking, this last lemma says that if the utility of Y falls between those of X_1 and X_2, then these latter two events will contain subevents that have the same utilities and also have the right probabilities to make it the case that

$$V(Y) = P(X_1{}^*/X_1{}^* \vee X_2{}^*)V(X_1) + P(X_2{}^*/X_1{}^* \vee X_2{}^*)V(X_2)$$

or, equivalently, that

$$\frac{V(Y) - V(X_2)}{V(X_1) - V(X_2)} = P(X_1{}^*/X_1{}^* \vee X_2{}^*)$$

This "solvability" property is crucial to Bolker's construction and a number of the results that will follow. Readers should make sure that they understand it before proceeding.

The next lemma shows that the Villegas probability P^* agrees with each Jeffrey/Bolker probability P_λ on every $\approx$-equivalence class in Ω.

Lemma 4.5e. *Suppose $(.>., .\geq.)$ and $(>, \geq)$ are EDT-sufficient. Let (P, V) be some expected utility representation of $(>, \geq)$ whose existence is secured by Bolker's Theorem (Theorem 4.3a). Let P^* be the unique probability that represents $(.>., .\geq.)$ whose existence is secured by Villegas's Theorem. Finally, let $\Gamma_Z = \{X \in \Omega \sim \textbf{N}: X \approx Z\}$ be the $\approx$-equivalence class of some nonnull $Z \in \Omega$. Then, for any $X, Y \in \Gamma_Z$, $P(X) = P(Y)$ if and only if $P^*(X) = P^*(Y)$.*

(The proof of this lemma is technical without being illuminating, and it may be skipped without loss of continuity.)

Proof. Suppose $X, Y \in \Gamma_Z$ and assume that $(Z \vee \neg Z) \geq Z$.[20] Use Corollary 4.5c to write X as $X_1 \vee X_2$ where X_1 and X_2 are incompatible, $X_1 .=. X_2$, and $X_1 \approx X_2$. Do the same for Y, so that $Y = (Y_1 \vee Y_2)$ where Y_1 and Y_2 are incompatible, $Y_1 .=. Y_2$, and $Y_1 \approx Y_2$. We will show that $P(X_2) = P(Y_2)$ iff $P^*(X_2) = P^*(Y_2)$, and it will follow directly that $P(X) = P(Y)$ iff $P^*(X) = P^*(Y)$.

---

[19] This is a special case of his Lemma 3.2, p. 300.
[20] When $Z > (Z \vee \neg Z)$ the proof works in the same way with $\neg G$ substituted for G.

Let G be the nonnull proposition such that $G > (G \vee \neg G)$ whose existence is guaranteed by *EDT$_1$. Partition G by setting

$$G_1 = (X_1 \ \& \ Y_1 \ \& \ G)$$
$$G_2 = (X_2 \ \& \ Y_2 \ \& \ G)$$
$$G_3 = (\neg G_1 \ \& \ \neg G_1 \ \& \ G)$$

Some of these propositions may be inconsistent, but at least one of them must be nonnull and ranked above $G \vee \neg G$. (This is because G is nonnull and ranked above $G \vee \neg G$). For purposes of illustration, suppose that G_1 has this feature.[21] Then, we have a case in which $G_1 > X_2 \approx Y_2$. Since both $G_1 \vee X_2$ and $G_1 \vee Y_2$ are sure to fall between G_1 and X_2 in the preference ranking, it follows from Coherence that $X_2 .=. Y_2$ if and only if $(G_1 \vee X_2) \approx (G_1 \vee Y_2)$. Since (P, V) represents $(>, \geq)$ and P^* represents $(.>., .\geq.)$ this means that

($\$$) $P^*(X_2) = P^*(Y_2)$ if and only if $V(G_1 \vee X_2) = V(G_1 \vee Y_2)$

But, $V(G_1 \vee X_2) = V(G_1 \vee Y_2)$ exactly if

$$P(G_1/G_1 \vee X_2)V(G_1) + P(X_2/G_1 \vee X_2)V(X_2)$$
$$= P(G_1/G_1 \vee Y_2)V(G_1) + P(Y_2/G_1 \vee Y_2)V(Y_2)$$

Since $V(G_1) > V(X_1) = V(Y_2)$ this reduces to $P(X_2/G_1 \vee X_2) = P(Y_2/G_1 \vee Y_2)$, and thence to $P(X_2) = P(Y_2)$ by the laws of probability. Thus, we may rewrite ($\$$) as

($\$\$$) $P^*(X_2) = P^*(Y_2)$ if and only if $P(X_2) = P(Y_2)$

It follows directly that $P^*(X) = 2P^*(X_2) = 2P^*(Y_2) = P^*(Y)$ just in case $P(X) = 2P(X_2) = 2P(Y_2) = P(Y)$, which is the result we were seeking. This completes the proof. ∎

When combined with Lemma 4.5a, Lemma 4.5e entails that, for each nonnull $Z \in \Omega$, there is a positive number c_Z such that $P^* = c_Z P$ everywhere on Γ_Z. We thus have:

Corollary 4.5f. $P^*(Z)/P(Z) = P^*(X)/P(X)$ *for any* $X, Z \in \Omega \sim \mathbf{N}$ *such that* $X \approx Z$.

We can use this fact to establish the following pivotal result:

[21] When G_3 is ranked above $\mathbf{T}$, replace G_1 in the rest of the proof by G_3 or G_4. When G_2 alone is ranked above $\mathbf{T}$, replace G_1 by G_2 and X_2 and Y_2 by X_1 and Y_1.

Lemma 4.5g. *If X and Y are nonnull propositions in Ω, then*

$$V(Y)[P^*(X)/P(X) - 1] = V(X)[P^*(Y)/P(Y) - 1]$$

Proof: If one of the propositions, say X, is ranked with $\mathbf{T} = (G \vee \neg G)$, then $V(X) = V(\mathbf{T}) = 0$ and $P^*(X)/P(X) = P^*(\mathbf{T})/P(\mathbf{T}) = 1$ (by the previous corollary), thus making the result trivial. To handle the case where neither X nor Y is ranked with $\mathbf{T}$, we begin by establishing the lemma when $Y = \neg X$. The trick is to use $I(\neg X) = -I(X)$ and the fact that $[P^*(X) - P(X)] = [P(\neg X) - P^*(\neg X)]$ to obtain the identity $I(\neg X)[P^*(X) - P(X)] = I(X)[P^*(\neg X) - P(\neg X)]$, from which the desired result follows directly since I is the product of V and P.

Let X and Y be ranked above $\mathbf{T}$ and let $V(X) \geq V(Y)$. If $V(X) = V(Y)$, the lemma follows from the previous corollary since $P^*(X)/P(X) = P^*(Y)/P(Y)$. On the other hand, if $V(X) > V(Y)$, then we have a situation in which $V(X) > V(Y) > V(\neg X)$ where $P(X)$ and $P(\neg X)$ are nonzero. Use the "solvability" lemma to find A and B that entail X and $\neg X$, respectively, and are such that $V(A \vee B) = V(Y)$, $V(A) = V(X)$ and $V(B) = V(\neg X)$. It follows from Corollary 4.5f both that $P^*(X)/P(X) = P^*(A)/P(A)$ and that $P^*(Y)/P(Y) = P^*(A \vee B)/P(A \vee B)$. Hence, to show

$$V(Y)[P^*(X)/P(X) - 1] = V(X)[P^*(Y)/P(Y) - 1]$$

it is sufficient to establish

$$V(A \vee B)[P^*(A)/P(A) - 1] = V(A)[P^*(A \vee B)/P(A \vee B) - 1]$$

To see why this last equality must be true, let $k = [P^*(A)/P(A) - 1]$ and proceed as in the previous paragraph to obtain

$$\begin{aligned} V(B)k = V(\neg X)k &= V(\neg X)[P^*(X)/P(X) - 1] \\ &= V(X)[P^*(\neg X)/P(\neg X) - 1] \\ &= V(A)[P^*(B)/P(B) - 1] \end{aligned}$$

And, using this result (in the third line), we get

$$\begin{aligned} V(A \vee B)[P^*(A)/P(A) - 1] &= V(A \vee B)k \\ &= [P(A/A \vee B)V(A)k] + [P(B/A \vee B)V(B)k] \\ &= [P(A/A \vee B)V(A)k(P^*(A)/P(A) - 1)] \\ &\quad + [P(B/A \vee B)V(A)(P^*(B)/P(B) - 1)] \\ &= V(A)[P(A/A \vee B)(P^*(A) - P(A)/P(A))] \\ &\quad + [P(B/A \vee B)(P^*(B) - P(B)/P(B))] \end{aligned}$$

$$= V(A)[(P^*(A) - P(A)/P(A \lor B))$$
$$+ P^*(B) - P(B)/P(A \lor B)]$$
$$= V(A)[(P^*(A) + P^*(B)) - (P(A) + P(B))]/P(A \lor B)$$
$$= V(A)[P^*(A \lor B)/P(A \lor B) - 1]$$

So we have shown that the lemma holds when X and Y are nonnull propositions that are strictly preferred to **T**. This suffices to establish the lemma in the general case because when X and Y are ranked below **T** we can apply the result of the last paragraph to $\neg X$ and $\neg Y$ (which will be ranked above **T**) to get

$$V(\neg Y)[P^*(\neg X)/P^*(\neg X) - 1] = V(\neg X)[P^*(\neg Y)/P^*(\neg Y) - 1]$$

and it follows that $V(Y)[P^*(X)/P(X) - 1] = V(X)[P^*(Y)/P(Y) - 1]$. Likewise, whenever **T** falls strictly between X and Y in the preference ranking we can apply the results of the previous two paragraphs to X and $\neg Y$ (or to $\neg X$ and Y) to obtain the desired result. This proves Lemma 4.5g. ∎

We are now in a position to complete the proof of Theorem 4.5. The goal, recall, was to show that P*, the unique probability that represents $(.>., .\geq.)$, is the first element in one of the Bolker representations (P_λ, V_λ) for $(>, \geq)$. To do this we had to find a constant λ between $-1/\inf\{V(X): X \in \Omega\}$ and $-1/\sup\{V(X): X \in \Omega\}$ such that $P^*(X) = P_\lambda(X) = P(X)(\lambda V(X) + 1)$ whenever $P(X) > 0$. Now, it follows from Lemma 4.5g that

$$[P^*(X)/P(X) - 1]/V(X) = [P^*(Y)/P(Y) - 1]/V(Y)$$

for any nonnull propositions $X, Y \in \Omega$ that are not ranked with **T** by $(>, \geq)$. Thus, the quantity $\lambda = [P^*(X)/P(X) - 1]/V(X)$ is a *constant* when $P(X) > 0$ and $V(X) \neq V(\mathbf{T})$. To see that λ falls between $-1/\inf\{V(X): X \in \Omega\}$ and $-1/\sup\{V(X): X \in \Omega\}$ note first that it will be in this interval just in case $\lambda V(X) > -1$ for $P(X) > 0$. This inequality holds trivially when $V(X) = 0$. When $V(X) \neq 0$ we can write $\lambda V(X) = ([P^*(X)/P(X) - 1]/V(X))V(X) = P^*(X)/P(X) - 1$, and this last quantity must be greater than -1 since $P^*(X)$ and $P(X)$ are both nonnegative real numbers. Theorem 4.3b then entails that $(P^*, V_\lambda) = (P_\lambda, V_\lambda)$ represents $(>, \geq)$. And, since Villegas's Theorem ensures that P* is the only probability that represents $(.>., .\geq.)$, it follows that, up to an arbitrary choice of a unit and a zero point for measuring utility, (P^*, V_λ) is the only joint representation of $(.>., .\geq.)$ and $(>, \geq)$.

This concludes the demonstration of Theorem 4.5. In the next two chapters we shall see how it can be used to provide representation theorems for a wide class of decision theories.

5

Causal Decision Theory

Evidential decision theory requires agents to choose actions that provide them with good evidence for thinking that desirable outcomes will occur. While this is usually sound advice, there are decision problems in which the pursuit of good news can lead to less than optimal results. This happens when there is a *statistical* or *evidential* correlation between the agent's choices and the occurrence of certain desirable outcomes, but no *causal* connection between the two. When the evidential and the causal import of actions diverge in this way, the evidential theory tells decision makers to put the pursuit of good news ahead of the pursuit of good results. Many philosophers, I among them, see this as a mistake. Rational agents choose acts on the basis of their *causal efficacy*, not their auspiciousness; they act to *bring about* good results even when doing so might betoken bad news.

In this chapter we examine *causal decision theory*, the version of expected utility theory that seeks to analyze the instrumental value of actions by appealing to their ability to causally promote desirable results. What we will see is that, even though evidential and causal decision theorists do disagree about the way in which we should evaluate actions, there is an underlying unity in their approaches to decision making. We will exploit this unity in subsequent chapters to prove a Bolker-style representation result for causal decision theory.

5.1 NEWCOMB PROBLEMS

The most famous decision problem in which evidential auspiciousness and causal efficacy diverge is Newcomb's Paradox.[1] Suppose there is a brilliant (and very rich) psychologist who knows you so well that he can predict your choices with a high degree of accuracy. One Monday as you are on the way to the bank he stops you, holds out a thousand dollar bill, and says: "You may take this if you like, but I must warn you that there is a catch. This past Friday I made a prediction about

---

[1] Newcomb's paradox was invented by William Newcomb. It was introduced to philosophers in Nozick (1969). The version presented here is due to J. H. Sobel.

what your decision would be. I deposited $1,000,000 into your bank account on that day if I thought you would refuse my offer, but I deposited nothing if I thought you would accept. The money is already either in the bank or not, and nothing you now do can change the fact. Do you want the extra $1,000?" You have seen the psychologist carry out this experiment on two hundred people, one hundred of whom took the cash and one hundred of whom did not, and he correctly forecast all but one choice. There is no magic in this. He does not, for instance, have a crystal ball that allows him to "foresee" what you choose. All his predictions were made solely on the basis of knowledge of facts about the history of the world up to Friday. He may know that you have a gene that predetermines your choice, or he may base his conclusions on a detailed study of your childhood, your responses to Rorschach tests, or whatever. The main point is that you now have no *causal influence* over what he did on Friday; his prediction is a fixed part of the fabric of the past. Do you want the money?

The decision you face can be represented by the following table:

	E = The psychologist predicted that you will take the money.	He predicted that you will refuse the money.
A = Take the money.	$1,000	$1,001,000
B = Refuse the money.	$0	$1,000,000

It is understood here that you regard E as very likely given A and very unlikely given B, that is, your subjective probability for the proposition $(A \ \& \ E) \vee (B \ \& \ \neg E)$ is nearly 1.

Evidential decision theory recommends refusing the money. Since the psychologist is likely to have correctly predicted your choice, whatever it turns out to be, it follows that doing A will give you strong evidence for thinking that E holds while doing B will give you strong evidence in favor of its negation. Taking the extra $1,000 is thus a reliable indication that the million is *not* in your account, whereas refusing it is an almost sure sign that the money is there. Since evidential decision theory recommends that you choose the act with the highest news value, it tells you to refuse the extra thousand to secure evidence of a million dollar payoff.

This amounts to treating Newcomb's Paradox like the "deterrence" example of the previous chapter. Here too it seems as if a dominance argument ought to apply. The outcomes in the "take the money" row are $1,000 greater than those in the "refuse the money" row, so taking the money looks like a sure thing. As we have seen, however, "sure-

147

thing" reasoning of this sort is legitimate only when the probabilities of states are independent of the decision maker's choice of an action, and that is not the case here. One of the consequences of taking the money is that you will *learn* that you take it, and this gives you evidence for thinking that, on Friday, the psychologist predicted that you would take it. Taking the money thus increases your confidence in *E*. Conversely, by refusing the money you cause yourself to learn that you will refuse it, thus decreasing your level of confidence in *E*. Hence, proponents of evidential decision theory argue, the dominance argument does not apply to Newcomb's Paradox because in it the probabilities of states depend on what act is performed.[2]

There are other decision problems that have this character. Suppose that you are playing the following (one shot) Prisoner's Dilemma game[3]

	Opponent cooperates.	Opponent defects.
You cooperate.	$8, $8	$0, $10
You defect.	$10, $0	$2, $2

with an opponent who is very much like you, for example, one who uses the same decision rule you use, but who will be making her choice in isolation. In such a situation, you might view your choice as a reliable indicator of your opponent's choice since your situations are symmetric. If so, evidential decision theory tells you to cooperate since this provides you with evidence for thinking that your opponent will also cooperate, which would be good news since you make an extra $8 if he does cooperate. This kind of reason seems to persuade many intelligent people.[4]

Another sort of case has to do with "the paradox of voter turnout," discussed in Chapter 2. The paradox, recall, was that people turn out to vote in large elections even though the probability of their vote's making any difference is so small that they should not think it worthwhile to cast even if the costs of voting are minuscule. There are surely many reasons why voters turn out in large elections: Voting may have symbolic value for many voters, others might be mistaken about the

---

[2] As David Lewis points out (1981, pp. 9–10), the probability of a correct prediction does not need to be terribly high for evidential decision theory to recommend refusing the money. In fact, if we use Bernoulli's logarithmic scale $u(\$x) = \ln(1 + x)$ for the utility of money, then refusing is good news when the probability of *E* given *A* is as small as 0.68 and the probability of $\neg E$ given *B* is anything larger than 0.68.

[3] Your payments come first; your opponent's are second.

[4] See, for example, Hofstadter (1983) and Horgan (1981) and (1985).

probabilities, some may think of it as a moral obligation, some people may like standing in lines, and so on. But, aside from all this, there may be evidentialist considerations that come into play. When asked why they vote people often respond by saying something like "My candidate can't win if people like me don't vote." Taken literally this is a *non sequitur*. It invites the question "Sure, but why did *you* come out to vote given that your voting is not going to affect the probability of 'people like you' voting to any measurable degree?" The voter's response only seems relevant if we focus on the phrase "people like me" and take her to be saying something like "My voting is good news because it indicates that people who support my candidate are likely to turn out." Once again, we have a person choosing an action on the basis of what it portends rather than what it causally promotes.

The structure of all these cases is the same. An agent faces a choice between two acts, A and B, where A provides her with much stronger *evidence* than B does for thinking that the world is in some desirable state, but where B has better *causal consequences* than A. Evidentialist reasoning leads an agent to choose A because it gives her "good news." The agent is supposed to evaluate her acts in precisely the same way that she evaluates states of the world over which she has no control. She should ask, "How desirable would it be if I were to receive news that I had done A?" and the answer is to be determined in much the same way as for a question like, "How desirable would it be if I received news of a flood in Bangladesh?" In a deep sense, the theory views propositions that describe aspects of the world that the agent can control as being on a par with those that describe events that lie entirely outside her influence. The clearest statement of the point is due to Jeffrey:

> To the extent that acts can be realistically identified with propositions, [my] notion of preference is active as well as passive: it relates to acts as well as to news items. . . . [this] notion of [utility] is neutral regarding the active-passive distinction. If the agent is deliberating about performing act A or act B, and if A & B is impossible, there is no effective difference between asking whether he prefers A to B as a news item or as an act, for he makes the news.[5]

Proponents of evidential decision theory thus recognize no significant distinction between treating a proposition that one can make true as an instrument for changing the world or as a piece of information about the world. This is why they are committed to refusing the extra $1,000 in Newcomb's Paradox, cooperating in Prisoner's Dilemma with a Twin, and so on.

[5] Jeffrey (1965/1983, p. 84). Jeffrey no longer espouses this view.

Proponents of *causal decision theory*, like me, regard this as entirely wrongheaded. There is a distinction, we claim, between genuine action and mere "news making." Statements that describe acts are different in kind from other sorts of propositions simply because the actor has the power to make them true. With this power comes a kind of responsibility. An agent must, if rational, do what she can to *change* things for the better. This means that the type of utility appropriate for use in an account of rational choice should *not* be neutral with regard to the active-passive distinction. Acts are not mere pieces of information; they are our means of altering reality. The classic statement of this view is due to Allan Gibbard and William Harper:

> The "utility" of an act should be its expected efficacy in bringing about states of affairs the agent wants, not the degree to which news of the act ought to cheer the agent . . . It may sometimes be rational for an agent to choose an act *B* instead of an act *A*, even though he would welcome the news of *A* more than that of *B*. The news of an act may furnish evidence of [a result] which the act itself is known not to produce. In that case, though the agent indeed makes the news with his act, he does not make all the news his act bespeaks.[6]

The point here is that rational decision makers should choose actions on the basis of their *efficacy in bringing about desirable results* rather than their auspiciousness as harbingers of these results. Efficacy and auspiciousness often go together, of course, since most actions get to be good or bad news only by causally promoting good or bad things. In cases where causing and indicating come apart, however, the causal decision theorist maintains that it is the causal properties of the act, rather than its purely evidential features, that should serve as the guide to rational conduct.

Perhaps the best way to appreciate the difference between evidential and causal decision theory is by seeing how each account interprets the principle of Dominance. As we saw in the last chapter, the evidential decision theorist claims that dominance reasoning should be applied only in cases where the state propositions used to characterize a decision problem are *evidentially independent* of the problem's acts, so that the agent's degrees of confidence in the states would not change were she to learn how she would behave. This leads to

Dominance with Evidential Independence. Let $\{E_1, E_2, E_3, \ldots\}$ be a partition of events that the decision maker regards as *eviden-*

[6] Gibbard and Harper (1978, pp. 356–57).

tially independent of her choice between A and B. If she weakly prefers A to B given E_j for every E_j, then she should weakly prefer A to B. Moreover, if any of these preferences is strict (and the associated event is nonnull), then she should strictly prefer A to B.

The proponent of causal decision theory, in contrast, insists that it is the causal, rather than the purely evidential, effects of the agent's acts that matter. Accordingly, they advocate

Dominance with Causal Independence. Let $\{E_1, E_2, E_3, \dots\}$ be a partition of events that the decision maker regards as *causally independent* of her choice between A and B. If she weakly prefers A to B given E_j for every E_j, then she should weakly prefer A to B. Moreover, if any of these preferences is strict (and the associated event is nonnull), then she should strictly prefer A to B.

This latter requirement is weaker than the former because an agent will never see her acts as causally relevant to the truth-values of states unless she also regards them as evidentially relevant – causation implies correlation – but the converse does not hold. As Newcomb-type cases illustrate, an agent can see evidential connections between actions and states even when she thinks the occurrence of those states is causally independent of what she does. Thus, proponents of the two decision theories will disagree about the appropriateness of employing Dominance in Newcomb's Paradox and similar problems. Causal decision theorists feel that the principle does legitimately apply, and thus they recommend taking the $1,000 from Newcomb's psychologist, defecting in Prisoner's Dilemma with a Twin, and staying home in large elections (if one's only reason for voting is to help one's candidate win). Evidential decision theorists see these as misuses of dominance reasoning, which they believe should be applied only when the agent's acts are both *causally* and *evidentially* irrelevant to the probabilities of states.

I am convinced that causal decision theory has it right here. While I am not going to rehearse the many arguments that have been offered on this score, it may be worthwhile to address the argument that proponents of refusing the money seem to find most convincing: the "If you're so smart why ain't you rich?" objection. I will deal with this objection in the context of the Newcomb Paradox, but my remarks apply, mutatis mutandis, to all other cases. A clear statement of the objection has been given by Robert Sugden. With respect to the dominance argument for taking the $1,000, he writes that

> Even if this argument is correct, we are left with the suggestion that [causal decision] theory may be in some sense self-defeating. Imagine two people, irrational Irene and rational Rachel, who go through the experiment. Irene is convinced by the [evidentialist] argument. She [refuses the money] and wins $1 million. Rachel is convinced by the [causalist] argument. She [takes the money] and wins $1,000. Rachel then asks Irene why she didn't [take the extra thousand]; surely Irene can see that she has just thrown away $1,000. Irene has an obvious reply: "If you're so smart why ain't you rich?"
>
> This reply deserves to be taken seriously. Both [evidentialist] and [causalist] arguments are based on an instrumental conception of rationality. It is taken as a given that more money is more desirable than less, and that rationality is about choosing the best means for reaching the most desired outcomes. The relevant difference between Irene and Rachel is that they reason in different ways. As a result of this difference, Irene finishes up with $1 million and Rachel with $1,000. Irene's mode of reasoning has been more successful measured against the common criterion of success. So, are we entitled to conclude that, nevertheless, it is Rachel who is rational?[7]

The wonderful thing about this passage is how clear it makes Irene's fallacy (although this is not what Sugden intended). In the last two lines of the first paragraph Rachel asks Irene why she (Irene) didn't take the extra thousand. Irene replies by asking Rachel why she (Rachel) isn't rich. This answer is, as lawyers say, "nonresponsive." Irene has replied to Rachel's question with an ad hominem that changes the subject. "I know why *I'm* not rich, Irene," Rachel should respond; "my question had to do with why didn't *you* take the $1,000." Irene might be tempted to answer that she didn't take the money because that would have placed her in Rachel's situation, thereby leaving her poor. But this is not the case. It is part of the definition of a Newcomb problem that the decision maker must believe that what she does will *not* affect what the psychologist has predicted. Letting M stand for "The $1,000,000 is in the bank," A be taking the money, and B be refusing it, this requires Irene to assign a high subjective probability to both of the following:

$$M \supset [(A \ \Box\!\!\rightarrow M) \ \& \ (B \ \Box\!\!\rightarrow M)]$$ If the $1,000,000 is in the bank, then it would still be there were I to do A and it would still be there were I to do B.

---

[7] Hargreaves Heap et al. (1992, p. 342). Sugden leaves this question rhetorical.

$$\neg M \supset [(A \,\square\!\!\rightarrow \neg M) \;\&\; (B \,\square\!\!\rightarrow \neg M)]$$ If the \$1,000,000 is not in the bank, then it would not be there were I to do A and it would not be there were I to do B.

If Irene really does believe these things, then she can give no good answer to Rachel's question since, having gotten the \$1,000,000, she must believe that she would have gotten it whatever she did, and thus that she would have done better had she taken the \$1,000. So, while she may feel superior to Rachel for having won the million, Irene must admit that her choice was not a wise one when compared to *her own* alternatives. The "If you're so smart why ain't you rich?" defense does nothing to let Irene off the hook; she made an irrational choice that cost her \$1,000.

But what of Rachel? Isn't this question still relevant to her? Ad hominems can, after all, hit their mark. Not this one. Rachel has a perfectly good answer to the "Why ain't you rich?" question. "I am not rich," she will say, "because I am not the kind of person the psychologist thinks will refuse the money. I'm just not like you, Irene. Given that I know that I am the type who takes the money, and given that the psychologist knows that I am this type, it was reasonable of me to think that the \$1,000,000 was not in my account. The \$1,000 was the most I was going to get no matter what I did. So the only reasonable thing for me to do was to take it."

Irene may want to press the point here by asking, "But don't you wish you were like me, Rachel? Don't you wish that you were the refusing type?" There is a tendency to think that Rachel, a committed causal decision theorist, must answer this question in the negative, which seems obviously wrong (given that being like Irene would have made her rich). This is not the case. Rachel can and should admit that she *does* wish she were more like Irene. "It would have been better for me," she might concede, "had I been the refusing type." At this point Irene will exclaim, "You've admitted it! It wasn't so smart to take the money after all." Unfortunately for Irene, her conclusion does not follow from Rachel's premise. Rachel will patiently explain that wishing to be a refuser in a Newcomb problem is not inconsistent with thinking that one should take the \$1,000 *whatever type one is*. When Rachel wishes she was Irene's type she is wishing *for Irene's options*, not sanctioning her choice. Rational Rachel recognizes that, whether she is the type that was predicted (on Friday) to take the money or the type that was predicted to refuse it, there is nothing she can do now to alter her type. She thus has no reason to pass up the extra $1,000. While a person who knows she will face (has faced) a Newcomb problem might wish that she were (had been) the type that the psychologist labels a refuser, this wish does not provide a reason for *being* a refuser. It might provide a reason to try (before Friday) to change her type *if she thinks this might affect the psychologist's prediction*, but it gives her no reason for doing anything other than taking the money once she comes to believe that she will be unable to influence what the psychologist does.

The response I just put in Rachel's mouth has her evaluating acts on the basis of *both* their auspiciousness and their efficacy. When she wishes she would refuse the extra $1,000 she is wishing for good news about herself, and when she decides not to take it she is deciding on the basis of efficacy. There is nothing wrong with simultaneously evaluating acts in both these ways. It should not be any part of causal decision theory to deny that acts have auspiciousness values or that these play a central role in our thinking about the desirability of prospects. Quite to the contrary, it turns out to be quite useful for causal decision theorists to have news values around since they help to explain why people so often feel "torn" when thinking about what to do in Newcomb-type problems. The deliberative tension is the result of auspiciousness pulling one way and efficacy pulling the other. The only thing we causal decision theorists are committed to is that efficacy should always win out in this tug-of-war when the issue is one of deciding how to act.

5.2 RATIFICATIONISM

While some proponents of auspiciousness maximization have tried to argue that turning down the extra $1,000 in Newcomb's problem is the rational act,[8] most now concede that it is not. Still, the promise of an analysis of rational choice free of causal entanglements remains appealing, and a number of writers have sought to modify the evidential theory so that it endorses taking the extra $1,000 without appealing to any unreconstructed causal judgments. The most influential of these attempts are the "tickle defense" of Ellery Eells, Richard Jeffrey's doctrine of *ratificationism*, and Huw Price's "agent probability" solution.[9]

All three approaches seek to provide an evidentialist solution to Newcomb-type problems by arguing that, contrary to first impres-

---

[8] See, for example, Horgan (1981).
[9] Eells (1982), Jeffrey (1965/1983), Price (1986, 1993).

sions, a rational agent will regard the states of such decisions as evidentially independent of her acts *at the time she chooses them* because at that instant she will have evidence that "screens off" any purely statistical correlations that obtain between her acts and the world's state. The notion of "screening off" comes from Hans Reichenbach's famous analysis of the concept of a common cause.[10] Reichenbach observed that collateral effects of a common cause are usually positively correlated with one another, but that knowledge of the cause tends to break the connection. Accordingly, he postulated that it is necessary (and sufficient) for C to be a common cause of E and E^* that knowledge of C should *screen off* the correlation between E and E^* in the statistical sense that the probabilities of E and E^* should be independent conditional on knowledge of C. In terms of subjective probabilities (of which Reichenbach was no great friend), we can put the point like this:

Reichenbach's Principle. An agent with subjective probability P regards C as a common cause of E and E^* only if (i) E and E^* are *not* statistically independent relative to P, so that $P(E \& E^*) \neq P(E)P(E^*)$, but (ii) E and E^* *are* statistically independent relative to P conditioned on C, so that $P(E \& E^*/C) = P(E/C)P(E^*/C)$.

For a simple application of this idea consider the high correlation between a person's having a Brooklyn accent and her having crossed the Brooklyn Bridge a large number of times. This is obviously not a case of direct causation; the two are joint effects of having lived in Brooklyn during one's early years. The existence of this common cause will be reflected in our beliefs, according to Reichenbach's Principle, by the independence of the two effects conditional on it. Once we know that the person hails from Brooklyn, the fact that she has a Brooklyn accent ceases to be (added) evidence for thinking that she has crossed the Brooklyn Bridge a great number of times, and the fact that she has crossed the Brooklyn Bridge a great number of times ceases to be (added) evidence for thinking that she has a Brooklyn accent.

We now know that Reichenbach's analysis is inadequate; there are "interactive forks" in which C causes both E and E^* even though $P(E \& E^*/C) > P(E/C)P(E^*/C)$.[11] There are even cases in which C screens off the statistical correlation between E and E^* when E causes E^*. Despite these failures, the notion of "screening off" remains a

[10] Reichenbach (1959, chapter 3).   [11] See Salmon (1984) and van Fraassen (1982).

useful tool in the philosophical study of causation; while lack of correlation after screening is not an *infallible* indicator of the lack of a direct causal connection between E and E^*, it often provides good *evidence* for this conclusion in garden-variety cases. This is relevant in the present context because Eells, Price, and Jeffrey all want to suggest, in different ways, that the lack of a causal connection between acts and states in Newcomb problems will typically be reflected in the decision maker's conditional beliefs in roughly the way that Reichenbach's Principle suggests.

Eells and Jeffrey both claim that a decision maker's *ability to anticipate her own choices* typically screens off any purely evidential import her acts might possess. Prior to performing an act, they maintain, a rational agent will always have factored in her decision to perform it, and the act itself will consequently cease to be a piece of evidence for her. Letting dA denote the *decision to do A*, the idea comes to this:

Screening. The decision to perform A screens off any purely evidential correlations between acts and states of the world, so that

$$P(E/B \ \& \ dA) = P(E/dA)$$

holds for all acts B and events E.

To ensure that these conditional probabilities are well defined we must assume that the agent always assigns a positive probability to the prospect that she will fail to carry out any given decision because of a "trembling hand," a lack of nerve, or other factors beyond her control, so that $P(B \ \& \ dA)$, though generally small unless $B = A$, is never zero. In other words, for *any* two actions the agent should assign some credence to the possibility that she will decide to do one of them but wind up doing the other.

To see what happens when Screening is introduced into the Newcomb problem, imagine that at some point during your deliberations you become convinced that you will decide to refuse the money, so that your degree of belief for dB moves close to 1. Since you are likely to carry out whatever decision you make, your degree of belief for B approaches 1 as well. This gives you strong evidence for thinking that the $1,000,000 is in the bank. Indeed, in the presence of Screening, it gives you evidence for thinking that the money is in the bank *no matter what you do*, so that $P(E/A \ \& \ dB)$ and $P(E/B \ \& \ dB)$ are *both* near 1. Thus, once you have *decided* on B, and moved from P to the revised probability P_{dB}, the events E and $\neg E$ become evidentially independent of your choice of an act. This makes it possible for you to apply the evidential dominance principle to conclude that taking the money is your most auspicious option. Hence, given a *decision* to refuse the money the only rational *act* is to take it. By the same token, if you become convinced that you will take the money, Screening will again make it reasonable for you to apply Dominance to justify taking it. Therefore, no matter how auspicious refusing the money might seem *before* you make up your mind about what to do, taking it is sure to look less auspicious *afterward*, and this will be true no matter what you decide. As long as Screening holds it is legitimate to conclude that *taking the money is the only rational course of action for an auspiciousness maximizer who knows what she has decided.*

Jeffrey and Eells propose to use this basic asymmetry, between what one decides and what one does, to argue that taking the extra $1,000 in Newcomb's problem is the only rational course of action for an auspiciousness maximizer. The case has already been made for an agent who already knows what she will decide. But what of those who have yet to make up their minds, those for whom neither P(dA) nor P(dB) is close to 1? How should they choose? Here the Eells/Jeffrey approach faces a serious challenge. If your subjective probabilities for dA and dB are far from 1, then the fact that B is better news than A should give you a reason to do B according to evidential decision theory. But if you take this reason to heart and decide to do B, then you will rue your decision because the news value of B is less than that of A given dB. This suggests that adhering to evidential decision theory makes preferences among acts unstable in a rather troubling way. It forces one into a position in which deciding on one course of action makes another look superior. This sort of "cognitive dissonance" between one's decisions and one's desires about having them executed is clearly something to be avoided.

While Eells and Jeffrey circumvent this problem in somewhat different ways, the similarities between their approaches are more striking than the differences. I will focus on Jeffrey's solution, but most of what I say applies to Eells's work as well. Jeffrey solves the problem by denying that one should always seek to maximize news value as one *currently* estimates it. If you are savvy, he argues, you will realize that any choice you make is likely to change some of your beliefs, thereby altering some of your desires. Your aim, therefore, should be to maximize news value not as you *now* estimate it, but as you *now* expect yourself to estimate it after having made a decision. You should "choose for the person you expect to be when you have chosen."[12] You do this by maximizing auspiciousness relative to the sub-

[12] Jeffrey (1965/1983, p. 16).

jective probability you expect yourself to have after having come to a firm decision about what to do. Jeffrey was thus led to posit the following as a necessary condition on rational choices:

The Maxim of Evidential Ratifiability. An agent cannot rationally choose to perform an act A unless A is *ratifiable* in the sense that she regards A & dA as better news than B & dA for every other act B under consideration.

This requires you to ignore your *current* views about the relative auspiciousness of taking the money verses refusing it in Newcomb's problem. Instead, you are supposed to maximize *future* auspiciousness by making choices that you can now ratify. "Ratificationism," as Jeffrey puts it, "requires performance of the chosen act, A, to have at least as high an estimated desirability as any of the alternative per-formances *on the hypothesis that one's final decision will be to perform A.*"[13]

The combination of the Maxim of Ratifiability and Screening secures the correct answer to Newcomb's Paradox without leaving the confines of evidential decision theory. In the presence of Screening deciding to take the money and then taking it are sure to make better news than deciding to take it and not taking it. Likewise, deciding not to take the money and then taking it are sure to make better news than deciding not to take it and not taking it. Given this, the ratifiability principle sanctions taking the money as the only reason-able choice because it is the only act that will maximizes news value *on the supposition that it is decided upon.* This provides an appropriately evidentialist rationale for taking the money, provided, of course, that the Maxim of Ratifiability and Screening are true.

While I think that the Maxim is true, Screening, unfortunately, is not. As Bas van Fraassen noticed,[14] there are versions of the Newcomb problem in which the performance of an act provides more secure evidence for the presence of the $1,000,000 in the bank than the mere decision to perform it. Suppose, for example, that you are a bumbler who is often unable to carry through successfully on your decisions. Imagine further that the psychologist is able to predict reliably not only your decision, but whether you will be able to execute it. Finally, suppose that he sets up the payments as follows:

---

[13] Jeffrey (1965/1983, p. 18).
[14] See Jeffrey (1965/1983, p. 20) for details.

158

	A & dA	B & dA	A & dB	B & dB
A = Take the money.	$1,000	$1,001,000	$1,000	$1,001,000
B = Refuse the money.	$0	$1,000,000	$0	$1,000,000

The relevant probabilities for these outcomes, we might imagine, are as follows:

Probabilities				
	A & dA	B & dA	A & dB	B & dB
P(•/A & dA)	0.8	0.1	0.09	0.01
P(•/A & dB)	0.09	0.01	0.8	0.1
P(•/B & dA)	0.1	0.8	0.01	0.09
P(•/B & dB)	0.01	0.09	0.1	0.8

These values reflect your belief that the psychologist will be right a high proportion of the time, about both the act decided upon and the act chosen. The reader may verify that with these probabilities, and any reasonable utility for money, refusing the extra $1,000 turns out to be the only ratifiable act. So Jeffrey's strategy does not provide a fully satisfactory evidentialist rationale for choosing the efficacious act in Newcomb problems. Jeffrey has admitted as much, writing that "ratifiability is not a completely reliable guide to choiceworthiness."[15]

Huw Price has suggested a somewhat different evidentialist rationale for taking the money in Newcomb's problem.[16] Price maintains that an agent who genuinely regards herself as free to act "cannot take the contemplated act to be probabilistically relevant to its causes."[17] In a later paper, he appeals to the authority of Ramsey, whom he portrays as holding the view that

> from the agent's point of view contemplated actions are always considered to be *sui generis*, uncaused by external factors. As [Ramsey] puts it, "my present action is an ultimate and the only contingency." I think this amounts to the view that free actions are treated as probabilistically independent of everything except their effects.[18]

The idea here is that a decision maker who takes herself to be a genuine agent is going to adopt a special epistemic perspective, encoded in a special "agent probability," relative to which her choice of

[15] Jeffrey (1965/1983, p. 20).   [16] Price (1986, 1991, 1993).
[17] Price (1986, p. 195).   [18] Price (1993, p. 261).

an act becomes evidentially irrelevant to her beliefs about the state of the world. This "agent probability" will *not* coincide with her ordinary subjective probability in Newcomb-type problems[19] because, Price claims, a decision maker will be unable simultaneously to regard herself as a free agent and yet recognize that there is a strong statistical correlation between what she does and what she has been predicted to do. Rationality, he says, requires auspiciousness maximization relative to the beliefs one holds *qua agent*, rather than one's ordinary beliefs. These special beliefs will be represented by "agent probabilities" that make states evidentially independent of acts even if one's ordinary subjective probabilities do not.

There is something important in this idea that a decision maker must evaluate her acts from a special epistemic perspective, and that her preferences go by auspiciousness when she does. This having been said, it is important to realize that such a view only counts as a version of *evidential* decision theory if the conditional "agent" probabilities of states given acts reflect meaningful *evidential* connections from the decision maker's point of view. The fact that the news value of taking the extra $1,000 is higher than the news value of refusing it relative to some *probability or other* is not surprising. Indeed, as Christopher Hitchcock has noted, it is possible to obtain this result within the context of a *causal* decision theory.[20] What Price needs to show is that his "agent probabilities" are both evidential in some appropriate sense and that they let V maximizers take the money in Newcomb problems. It is hard to see how this can happen. All it takes to make a Newcomb problem is a noncausal correlation between acts and states. However, since our evidence is always evidence about observed correlations, our conditional probabilities, taken by themselves and unsupplemented by explicitly causal information, are never going to account fully for differences between causal connections and mere noncausal correlations. This, as I see it, is little more than an application of Hume's famous point that evidence about "constant conjunctions" among events will never be sufficient to reveal the "necessary connections" that obtain between cause and effect.

In the end, I do not believe that there is any acceptable general way to get evidential decision theory to yield the correct answers in the Newcomb Paradox and similar cases. To get the right answers in these

[19] In his (1986) Price seems to have thought that these probabilities would be the same if the decision maker takes all her evidence into account, including the fact that she believes that her choice is positively correlated with the decision maker's prediction. This turns out not to be untenable. See Sobel (1994, pp. 39–40).
[20] Hitchcock (1996).

decision problems one needs to appeal explicitly to the decision maker's beliefs *about causal relations*. In saying this I am placing myself squarely in the camp of the causal decision theorists who believe, with David Lewis, that

noncausal decision theory gives the wrong answer [in Newcomb problems]. It commends the irrational policy of managing the news so as to get good news about matters over which you have control . . . We need an improved decision theory, more sensitive to causal distinctions. Noncausal decision theory will do when the causal relations are right for it, as they very often are, but even then the full story is causal.[21]

Thus, if Price's "agent probabilities" are going to be able to capture the "causal story" of rational decision making adequately they must reflect something more than purely evidential relationships between actions and states; they must reflect the agent's beliefs about what her actions are capable of causing.

5.3 CAUSAL DECISION THEORY

Causal decision theory seeks to provide a rigorous formal analysis of the idea that a rational decision maker should evaluate her potential actions solely on the basis of their ability to cause desirable outcomes. Abstractly speaking, a *causal expected utility* is a function of the form

$$\textbf{CDT.} \quad U(A) = \sum_s P(S \setminus A)u(A \ \& \ S)$$

where the probability function $P(\bullet \backslash A)$ provides a measure of the agent's estimates of A's "causal tendencies." The quantity $U(A)$, hereafter A's *efficacy value*, gauges the extent to which performing A can be expected to *bring about* desirable or undesirable outcomes.

The key to the equation CDT is the probability $P(\bullet \backslash A)$. This quantity has been interpreted in a variety of ways in the literature, but, stepping back a bit from the fray, we can say that the common ground among causal decision theorists is that $P(\bullet \backslash A)$'s values should reflect a decision maker's judgments about her ability to *causally influence* events in the world by doing A. I will call it her *causal probability for A*. More generally, a *causal probability function* is a two-place operation $P(X \backslash Y) = P^Y(X)$ defined on (at least) $\Omega \times \Omega(\textbf{A})$ such that

[21] Lewis (1981, p. 5).

161

- Each $P(\bullet\backslash Y)$ is a probability defined over propositions in Ω.
- $P(Y\backslash Y) = 1$ for all Y for which $P(\bullet\backslash Y)$ is defined.
- $P(X\backslash Y) > P(X\backslash Z)$ just in case the agent judges that either Y will causally promote X's occurrence more strongly than Z will or Y will causally inhibit X's occurrence less strongly than Z will.

In light of this last clause it makes sense to say that the agent views Y a *promoting cause* of X just in case $P(X\backslash Y)$ is larger than $P(X\backslash \neg Y)$, and that she views $\neg Y$ as an *inhibiting cause* in these same circumstances. This justifies all of the following terminology:

- Y is *positively casually relevant* to X exactly if $P(X\backslash Y) > P(X\backslash \neg Y)$.
- Y is *negatively casually relevant* to X exactly if $P(X\backslash Y) < P(X\backslash \neg Y)$.
- Y is *casually relevant* to X if and only if $P(X\backslash Y) \neq P(X\backslash \neg Y)$.
- X is *causally independent* of Y if and only if $P(X\backslash Y) = P(X\backslash \neg Y)$.
- Y is a *more efficient cause* of X than Z is if and only if $P(X\backslash Y) > P(X\backslash Z)$.

These equivalences should look familiar since all have well-known evidential analogues. If one replaces $P(X\backslash Y)$ by $P(X/Y)$ and substitutes "evidentially" for "causally," and "better evidence for" for "more efficient cause of," then one gets the standard Bayesian characterization of comparative evidential relationships in terms of conditional probabilities. The front slash "\" is thus a kind of causal analogue of the evidential backslash "/."

As in the evidential case, it is important to emphasize that the expressions here have to do with judgments of *relative* causal efficacy. $P(X\backslash Y)$ does *not* measure Y's influence over X in any absolute sense. The fact that it is high (low) does not mean that the agent believes that Y will (or won't) cause X. Rather, $P(X\backslash Y)$ will be high *either* when she thinks that Y will cause X or when she thinks that X is likely to hold *whether or not Y does*. If, for example, X says that Cleveland will get snow next January and Y is the proposition that I order salad for lunch today, then my value for $P(X\backslash Y)$ is quite high even though I do not think my eating habits can affect the weather in Ohio. This lack of causal power is indicated by the fact that, for me, $P(X\backslash Y) = P(X\backslash \neg Y)$.[22]

The challenge for causal decision theorists is to define $P(\bullet\backslash A)$ in a

---

[22] The evidential analogue is instructive here as well. My conditional probability for X given Y is very high even though I do not regard Y as evidence for X. This is because, for me, $P(X/Y)$ and $P(X/\neg Y)$ are equal and both close to 1.

way that ensures that it captures relationships of causal dependence and causal independence in the manner described. Proponents of the theory tend not to put the issue to themselves in this abstract way; since each has his or her own candidate for $P(\bullet \backslash A)$, the general point tends to get lost. For our purposes, though, it is crucial to focus on the theme rather than the variations. We should always think of any version of causal decision theory as involving two distinct elements: (1) the claim that rational agents should evaluate their potential actions with an eye toward maximizing a utility function of the form CDT, and (2) some substantive proposal for interpreting $P(X \backslash Y)$. Claim (1) is an axiological thesis that has to do with how people should desire things, while (2) is an epistemic and metaphysical claim that concerns our understanding and knowledge of causal relations. There is no *a priori* reason to suppose that an adequate justification of the first must await a satisfactory rationale for the second. Indeed, we will see that one can gain a deep understanding of the foundations of causal decision theory by pursuing the first question more or less in isolation from the second. Before doing so, however, let's take a moment to see how casual decision theorists have tried to address the second issue.

The one point on which all causal decision theorists agree is that $P(X \backslash A)$ is *not* $P(X/A)$: Evidential relevance isn't causal relevance; correlation isn't causation; indicating is not promoting. Beyond this negative thesis, most proponents of the theory favor what I'll call a *K*-partition account of causal probabilities.[23] Leaving aside intramural differences, the idea is that the efficacy values of acts are unconditional expected utilities (à la Savage) computed relative to a special partition *K* whose elements provide a maximally specific description of the ways in which things the agent cares about might depend on what she does. Following David Lewis, I will call *K*'s elements *dependency hypotheses*.[24] These hypotheses are supposed to have two properties that make them ideal for use in calculating causal expected utilities. First, the decision maker must regard them as causally independent of what she does. Second, they are supposed to contain enough information about causal relationships so that, for any fixed $K \in K$, differences in the subjective conditional probabilities of the form $P(E/A \ \& \ K) - P(E/B \ \& \ K)$ measure the extent to which the agent thinks that A is more likely than B to promote E, or less likely to inhibit it, *given the causal story told by K*.

[23] See Lewis (1981), Skyrms (1984), and Armendt (1986).
[24] Lewis (1981, p. 11).

The first requirement sanctions the use of the following analogue of Savage's equation to compute causal expected utilities:

$$\mathbf{CDT}_K \quad U(A) = \sum_K P(K)V(A \ \& \ K)$$

An act proposition's efficacy value is thus a weighted average of the *auspiciousness* of each of the ways in which it can be true when conjoined with dependency hypotheses in **K**. If an agent knows the full causal structure of the decision problem she faces, so that she assigns some specific $K \in \mathbf{K}$ subjective probability 1, then she will choose acts on the basis of their news value (since $V(A \ \& \ K) = V(A)$ when $P(K) = 1$). This represents the common ground between causal and evidential decision theorists; both agree that good news reliably indicates good results when all the causal facts are known (where this includes facts about what outcomes will occur whatever the decision maker does).

In trying to understand the **K**-partition approach it is helpful to use the fact that atoms of Ω are unalloyed goods to write the value of each act/state conjunction as $V(A \ \& \ K) = \sum_W P(W/A \ \& \ K)u(A \ \& \ K \ \& \ W) = \sum_W P(W/A \ \& \ K)u(W)$ where W ranges over the atoms of Ω, and then to reformulate CDT_k as

$$\mathbf{CDT}_K \quad U(A) = \sum_W \left[\sum_K P(K)P(W/A \ \& \ K)\right]u(W)$$

This expresses CDT_k as an instance of the general equation for causal expected utility CDT. The function $\sum_K P(K)P(\bullet/A \ \& \ K)$, the agent's *expectation* of $P(\bullet/A \ \& \ K)$ as K ranges over elements of **K**, will be a probability over Ω as long as $P(A) > 0$.[25] Proponents of the **K**-partition approach maintain that this is the way to measure A's "causal tendencies." They thus endorse the

K-expectation Definition of Causal Probability. If $P(A) > 0$, then

$$P(\bullet \setminus A) = \sum_K P(K)P(\bullet \setminus A \ \& \ K)$$

for some appropriate choice of a partition of dependency hypotheses **K**.

The agent is thus portrayed as believing that an act A is a promoting cause of an event E just in case she *expects* E to have a higher subjective probability conditional on $A \ \& \ K$ than conditional on $\neg A \ \& \ K$ where K varies over dependency hypotheses in **K**.

---

[25] The requirement that $P(A) > 0$ is inessential since one can suppose that these conditional probabilities are given by a Réyni–Popper measure. See Chapter 7.

164

The hard part is finding the right K. There are three ways to go. I will discuss them rather quickly and somewhat superficially because their differences are less important than their similarities in what follows. The first approach uses some version of the "probabilistic" account of causation pioneered by Patrick Suppes, and subsequently developed by Nancy Cartwright, Ellery Eells, Christopher Hitchcock, and a host of others.[26] Here one helps oneself to a primitive notion of *a relevant causal factor* and seeks to (nonreductively) explain an agent's causal judgments in terms of her subjective probabilities conditional on certain causal factors being held fixed. Intuitively, to say that one variable quantity f (which can be a proposition) is a *causally relevant factor* for another, g, means roughly that a complete causal explanation for g's having the value it does must make mention of the fact that f has the value it does. This is admittedly vague, but the underlying idea is clear enough in particular cases – for example, we all know that the number of cigarettes one smokes, the grams of fat in one's diet, the amount of exercise one gets, and one's genetic makeup are causal factors implicated in heart disease. In the context of a decision problem $D = (\Omega, O, S, A)$ a partition K is *causally homogeneous* relative to A and S just in case each of its elements provides a complete specification of the values of those quantities (expressible in Ω) that (i) the decision maker sees as causally relevant to the truth of propositions in A and S, but that (ii) are not judged to have any of the acts in A among their own causal factors. They are, as Skyrms puts it, "maximally specific specifications of the factors outside our influence [thus (ii)] that are causally relevant to the outcomes of our actions [thus (i)]."[27] Each cell in a causally homogeneous partition corresponds to one way of "holding fixed" the causal "background conditions" that the decision maker believes to be relevant to her acts and their outcomes. Relative to such a partition, the claim is, we can cash out the agent's beliefs about the causal consequences of her actions in probabilistic terms. She is supposed to see A as a more efficient cause of E than B is just in case she expects the conditional probability of E given A to be higher than the conditional probability of E given B *when all causal factors other than A and B, and whatever they might cause, are held fixed.* Given such a reading of causal beliefs, it is natural to suppose that expected utilities should be computed using CDT_k with a K partition that is causally homogeneous with respect to A and S.

---

[26] Essential references here are Suppes (1970), Cartwright (1979), Eells (1991), and Hitchcock (1993).

[27] Skyrms (1980, p. 133).

A second, closely related method for fixing **K** involves supposing that dependency hypotheses provide direct specifications of objective probabilities, or *chances*, as they are often called in the literature. Adherents of this approach imagine that each K includes a complete *theory of objective conditional chance*, so that it entails a statement of the form $CH(E/A) = x$ (read, "The chance of E conditional on A is x")[28] for each event E and act A, where these assignments collectively obey the laws of conditional probability. In decision making contexts, it generally makes sense to suppose that the relationship between conditional chance and subjective probability is governed by the

Principal Principle (conditional version).[29] *If P is any probability on Ω, if P(A & K) > 0, and if K entails that the chance of E conditional on A is x, then P(E/A & K) = P(E/A & CH(E/A) = x) = x.*

When $P(A) > 0$, this lets us write $P(\bullet\backslash A) = \Sigma_K P(K)P(\bullet/A \& K) = \Sigma_x x P(CH(\bullet/A) = x)$ where x ranges over $[0,1]$. $P(\bullet\backslash A)$ thus becomes the decision maker's *expectation of objective chance conditional on A*. It is not at all implausible to suppose that estimates of this sort underwrite many of our causal beliefs. When I tell you to quit smoking because it causes cancer I mean to convey something like this: "If you smoke you run a risk of contracting cancer that is higher than the risk you run if you do not smoke. I am not just saying something about my evidence, mind you; the fact is that you have a higher *objective* risk of contracting cancer if you smoke than if you do not, and this risk does not depend on what you, I, or anyone else thinks." Reflection on examples like this makes it plausible to think that a difference in estimates of conditional chances often indicates a difference in causal powers. If so, differences in a person's estimates of $CH(E/A)$ and $CH(E/B)$ should correspond to differences in her beliefs about the causal tendencies of A and B *vis-à-vis* E.

Proponents of the "chance" reading of dependency hypotheses come in two flavors. *Objectivists* maintain that there is a *single* **K** partition of conditional chance hypotheses that holds good for all

[28] Chances are always indexed to times. In decision making contexts, the time in question here is always the time at which the agent makes her choice. This will be left implicit throughout the discussion to save on subscripts.

[29] This principle is discussed in Lewis (1980) and (1994). The conditional version is due to Skyrms (1988) and is discussed in Gaifman (1988). There are certain restrictions on the Principal Principle that I have left out, namely, that it applies only when neither A nor K nor P contains "inadmissible information" that speaks directly to S's truth. In all reasonable decision-theoretic situations this will be the case.

agents. They claim, more precisely, that each sentence "CH(E/A) = x" expresses a definite proposition that does not vary from person to person, and that each rational subjective conditional probability function must obey the Principal Principle, so that P(E/A & CH(E/A) = x) = x for any reasonable probability P. In contrast, *subjectivists* regard conditional chances as "emergent" properties of the agent's opinions.[30] Under the right conditions, they say, there will be a set of hypotheses $\{C_x(E, A): E \in \Omega, A \in \boldsymbol{A}, x \in [0,1]\}$, *that depends on the agent's subjective probabilities and which varies from person to person*, such that the identity P(E/A & $C_x(E, A)$) = x holds for all $E \in \Omega$ and $A \in \boldsymbol{A}$.[31] The chance statement CH(E/A) = x can then be reductively identified with the $C_x(E, A)$, which makes the Principal Principle hold by definition, and the relevant $\boldsymbol{K}$ partition can be identified with maximally consistent conjunctions of these *agentrelative* chance propositions. The important thing, for present purposes, is that both objectivists and subjectivists agree that an agent's causal beliefs are reflected in her estimates of chances according to the rule that A is a more efficient cause of E than B is iff $\Sigma_x x$P(CH(E/A) = x) > $\Sigma_x x$P(CH(E/B) = x). They also agree that dependency hypotheses are maximally detailed specifications of conditional chances.

A third version of the $\boldsymbol{K}$ partition approach, which has its genesis in the work of Robert Stalnaker, Allan Gibbard, and William Harper, seeks to understand beliefs about causal tendencies in terms of *counterfactual dependence*. Here, a person's views about A's causal powers reveal themselves in her unconditional degrees of belief for subjunctive conditionals. She will regard A as promoting E exactly if P($A \;\square\rightarrow E$) > P($\neg A \;\square\rightarrow E$) and as inhibiting E if and only if P($A \;\square\rightarrow E$) < P($\neg A \;\square\rightarrow E$), and will regard E as causally independent of A when P($A \;\square\rightarrow E$) = P($\neg A \;\square\rightarrow E$). More generally, she will see A as more efficient than B as a cause of E exactly if P($A \;\square\rightarrow E$) > P($B \;\square\rightarrow E$).[32] When relative causal tendencies are understood in this way, we can replace P($\bullet \backslash A$) by P($A \;\square\rightarrow \bullet$) in CDT to obtain *Stalnaker's Equation*:

$$\mathbf{CDT}_{SE} \quad U(A) = \sum_W P(A \;\square\rightarrow W)u(W)$$

---

[30] The pioneering work here is de Finetti (1964, sect. 12.6).

[31] It is a tricky matter to define this set of propositions. This was accomplished by Brian Skyrms. See his (1984).

[32] To understand this inequality subtract P(($A \;\square\rightarrow E$) & ($B \;\square\rightarrow E$)) from both sides to get P(($A \;\square\rightarrow E$) & $\neg$($B \;\square\rightarrow E$)) > P($\neg$($A \;\square\rightarrow E$) & ($B \;\square\rightarrow E$)). In the terminology of Gibbard and Harper (1978) this says that the probability of E's being a *causal consequence of A as opposed to B* exceeds the probability of E's being a *consequence of B as opposed to A*.

In order to know that this function is well defined we must be certain that $P(A \,\square\!\!\rightarrow \bullet)$ obeys the axioms of probability. It turns out that it will, provided that "$\square\!\!\rightarrow$" is a conditional (see the axioms of Chapter 2) that satisfies

Conditional Contradiction. If φ and φ^* are contraries and χ is not a contradiction, then both $\chi \,\square\!\!\rightarrow \varphi$ and $\chi \,\square\!\!\rightarrow \varphi^*$ are also contraries.

Harmony. If χ and χ^* are contraries, and if neither $\chi \,\square\!\!\rightarrow \varphi$ nor $\chi^* \,\square\!\!\rightarrow \varphi^*$ is a logical falsehood, then $(\chi \,\square\!\!\rightarrow \varphi)$ & $(\chi^* \,\square\!\!\rightarrow \varphi^*)$ is not a logical falsehood.

Conditional Excluded Middle. $(\chi \,\square\!\!\rightarrow \varphi) \vee (\chi \,\square\!\!\rightarrow \neg\varphi)$ is a logical truth.

Everything but the last requirement is unproblematic. Under the semantics developed by Stalnaker and Lewis,[33] the truth-values of subjunctive conditionals are fixed in terms of a relation $\text{SIM}_\alpha(W, W^*)$, defined on pairs of possible worlds W and W^*, which signifies that W is at least as "close" as W^* is to the actual world α in terms of *relevant overall similarity*. SIM_α has the following formal properties:

- *Transitivity*: If $\text{SIM}_\alpha(W, W^*)$ and $\text{SIM}_\alpha(W^*, W^{**})$, then $\text{SIM}_\alpha(W, W^{**})$.
- *Connectedness*: $\text{SIM}_\alpha(W, W^*)$ or $\text{SIM}_\alpha(W^*, W)$ for all worlds W and W^*.
- *Centering*: $\text{SIM}_\alpha(\alpha, W)$ holds for every world W.

Relative to such a relation, one says that $(A \,\square\!\!\rightarrow X)$ is true either if A is logically false or if there exists a world in which A & X holds that is "more like" the actual world, according to SIM_α, than any $(A$ & $\neg X)$-world. The reader is invited to verify that this semantics for $\square\!\!\rightarrow$ ensures that it is a genuine conditional (i.e., that it obeys Cond_1–Cond_8 of Chapter 2), and that it satisfies both Conditional Contradiction and Harmony.

It does not, however, guarantee Conditional Excluded Middle since both $A \,\square\!\!\rightarrow X$ and $A \,\square\!\!\rightarrow \neg X$ can be false either when there exists a "circle" of worlds, all equally close to actuality, that contains both $(A$ & $X)$-worlds and $(A$ & $\neg X)$-worlds, or when there is an infinite sequence of $(A$ & $X)$-worlds $W_1, W_2, W_3, \dots$ and an infinite sequence

---

[33] Stalnaker (1968) and Lewis (1973). I am presenting a special case of their semantics that deals exclusively with similarity *with respect to the actual world*. The full semantics treats similarity with respect to an arbitrary world. Also, I am assuming that every possible world is "accessible" from the actual world, and that no world is as similar to the actual worlds as it is to itself (this is *Centering*). Lewis and Stalnaker give treatments in which these assumptions can be relaxed.

of (A & $\neg X$)-worlds W_1^*, W_2^*, W_3^*, ... such that W_j is always more like the actual world than W_j^* but W_{j+1}^* is always more like the actual world than W_j. In both these cases Conditional Excluded Middle fails.

The Stalnaker formula makes no sense without Conditional Excluded Middle since the function $P(A \;\square\!\!\rightarrow\; \bullet)$ is not additive unless $\neg(A \;\square\!\!\rightarrow\; E)$ and $A \;\square\!\!\rightarrow\; \neg E$ are equivalent. Stalnaker was able to secure this by imposing the following constraint on the semantics of subjunctive conditionals:

Stalnaker's Assumption. *For each contingent statement A, there is a unique possible world W_A in which A is true that is more like the actual world than any other world in which A is true, so that $SIM_a(W_A, W)$ holds for any A-world W.*

This makes $A \;\square\!\!\rightarrow\; X$ true if and only if X is true in W_A. Since the (unconditional) Law of the Excluded Middle will hold in W_A, this implies the truth of Conditional Excluded Middle. There are, as we shall see, some significant problems with Stalnaker's Assumption, but let's put them aside for the moment and see what happens when it is in place.

While the formal requirements presented in the last few paragraphs do ensure that $P(\bullet\backslash A) = P(A \;\square\!\!\rightarrow\; \bullet)$ is a probability, they do not guarantee that it is reasonable to use it to describe a decision maker's beliefs about causes and effects. For that, the conditional must be interpreted in a way that reflects causal connections; the statement $A \;\square\!\!\rightarrow\; E$ must be what David Lewis calls a "causal counterfactual."[34] As Lewis points out, similarity among possible worlds is a vague notion, and its vagueness can be resolved in different ways in different contexts. Causal counterfactuals are evaluated in terms of what Lewis calls the "standard resolution" of vagueness, as opposed to the "back-tracking" resolution:

We know that present conditions have their past causes. We can persuade ourselves ... that if the present were different then their past causes would have been different, else they would have caused the present to be what it actually is. Given such an argument – let us call it a *back-tracking argument* – we willingly grant that if the present were different, the past would be different too. ... [But] we very easily slip back into our usual sort of counterfactual reasoning, and implicitly assume that facts about earlier times are counterfactually independent of facts about later times ... We ordinarily resolve the vagueness of counterfactuals in such a way that counterfactual dependence is asymmetric (except in cases of time travel and the like). Under this standard

[34] Lewis (1981, p. 22).

resolution, back-tracking arguments are mistaken. . . . A counterfactual saying that the past would be different if the present were somehow different may come out true under the [back-tracking] resolution of its vagueness, but false under the standard resolution.[35]

The point, then, is that the subjunctive conditionals that are used in causal decision theory should be interpreted in such a way as to mirror the asymmetries inherent in causal relations. When evaluating $A \, \square \rightarrow E$ one looks for the "nearest" possible world in which A is true *among those worlds that agree with the actual world on matters of historical fact up to the moment of A's performance.*[36] In Newcomb's Paradox, for example, you are not allowed to change the psychologist's prediction by reasoning, in backtracking fashion, that if you were to take the money, then the psychologist would have predicted your taking it. Under a nonbacktracking construal of "similarity," the claim is, an agent will assign a higher probability to $A \, \square \rightarrow E$ than to $B \, \square \rightarrow E$ only if she thinks that A is more likely than B to promote E (or less likely to inhibit E). Thus, the subtext of the Stalnaker Equation should read, "where each conditional $A \, \square \rightarrow W$ is evaluated under the standard, causal interpretation of similarity among possible worlds."

We can bring the Stalnaker/Gibbard/Harper theory into line with the **K**-expectation approach by defining dependency hypotheses as (consistent) conjunctions of subjunctive conditionals of the form $K = \&_A(A \, \square \rightarrow W_{A,K})$ where A ranges over all the actions in **A**, and each $W_{A,K}$ is an outcome of the form $A \, \& \, S$ where $S \in$ **S**. Such a K provides a complete specification of the way in which the world's state depends counterfactually on the agent's actions. In the presence of Conditional Contradiction, Harmony, and Conditional Excluded Middle, the K's form a partition, and it is reasonable to assume that the decision maker will take them to be causally independent of her choices.[37] It makes sense, then, to apply the **K**-expectation definition of causal probability to each outcome W and act B to obtain

$$P(W \setminus B) = \sum_K P(\&_A(A \, \square \rightarrow W_{A,K}))P(W \, / \, B \, \& \, [\&_A(A \, \square \rightarrow W_{A,K})])$$

<hr/>

[35] Lewis (1979, pp. 456–57).

[36] Notice, however, that it is possible to reason to the contrapositive conclusion that the psychologist would only have predicted that you would take the money if you would in fact take it. Indeed, this is what it means to call him reliable.

[37] Whenever it seems as if the agent has the ability to alter the probability of some proposition of the form $K = \&_A(A \, \square \rightarrow S_{A,K})$ this will be because the **A** does not really contain all of her options. Specifically, it does not contain the action she would have to take to alter K's likelihood.

provided that $P(B) > 0$. As Gibbard and Harper showed,[38] the right-hand side of this identity reduces to $P(B \, \square\!\!\rightarrow W)$ under the assumptions about subjunctive conditionals that are in place, and this shows that the two ways of interpreting $P(W \backslash B)$ in terms of subjunctive conditional agree. Thus, CDT_k and Stalnaker's Equation yield the same results when dependency hypotheses have form $K = \&_A(A \, \square\!\!\rightarrow W_{A,K})$ provided that A has positive probability.

From a purely decision-theoretic perspective, the similarities among the "causally relevant factor," "conditional chance," and "counterfactual dependence" versions of the **K**-partition approach are more striking than their differences. A person's judgment about which of the three interpretations is correct will depend on her views about the metaphysics and epistemology of causation rather than her intuitions about how to make choices. Those who go in for probabilistic analyses of causation will favor the first alternative. Those who think causation should be cashed out in terms of objective chances will identify dependency hypotheses with complete specifications of conditional chances. And, if one thinks, as I do, that causation is best understood in subjunctive terms, then the Stalnaker/Gibbard/ Harper approach will suit one's fancy. Moreover, no matter which interpretation one likes, one will be inclined to suppose that the other two say essentially the same thing in different terms. Indeed, causal decision theorists have done a great deal of work aimed at minimizing the differences among their formulations.[39] (The "subjectivist's" reduction of conditional chance to rational degree of belief conditional on a special partition of events is a prime example of this.) David Lewis seems to have gotten it right when he wrote, "We causal decision theorists share one common idea, and differ mainly on matters of emphasis and formulation."[40] Skyrms and Harper have expressed the same sentiment, observing, "It can be argued that the various forms of causal decision theory are equivalent – that an adequate version of any one of [them] will be interdefinable with adequate versions of the others."[41] I think this is basically right; there is no great difference between any of the approaches we have discussed. This should not be terribly surprising. As philosophers of science have long been telling us, the notions of causation, chance, counterfactual dependence, similarity among worlds, and natural law form a constellation of interrelated concepts, any one of which can be used as a starting point for an

---

[38] For a proof that these readings of $P(\bullet \backslash B)$ agree see Gibbard and Harper (1978).
[39] The best work done along this line has been that of Brian Skyrms. See his (1984) and (1994) for relevant details.
[40] Lewis (1981, p. 5).   [41] Harper and Skyrms (1988, p. x).

analysis of the rest. Whatever the merits of starting in one place or another in metaphysics or epistemology, the causal decision theorist can adopt an attitude of benign indifference; she can do her work as long as the causal powers of actions can be analyzed in one or another of these ways.

5.4 SUBJUNCTIVE CONDITIONALS AND IMAGING

That having been said, I'd like to register my preference for a subjunctive reading of $P(\bullet \backslash A)$ and to say a bit more about how the view should be developed. I regard decision making as an essentially *subjunctive* endeavor; to evaluate an act, in the causally relevant sense, is to ask what *would* happen *were* it to be done. Given such an outlook, each causal probability $P(W \backslash A)$ that appears in CDT should capture the decision maker's views about *how likely W would be were A to be performed* (keep in mind, this need not be the same as how likely W is if A is performed). This suggests that $P(\bullet \backslash A)$ is best identified with $P(A \ \square \rightarrow \ \bullet)$, and that causal utilities should be computed by using Stalnaker's Equation. One might even propose it as a *constraint* on **K** partitions that $P(\bullet / A \ \& \ K) = P(A \ \square \rightarrow \ \bullet / K)$ should hold for all $K \in$ **K**, so as to ensure that $P(A \ \square \rightarrow \ \bullet) = \Sigma_K P(K) P(\bullet / A \ \& \ K)$ holds for all acts A.

Something like this is right, but there are subtleties that need to be taken into account. The problem is that Stalnaker's Equation presupposes Conditional Excluded Middle and Stalnaker's Assumption, and David Lewis has argued convincingly that both are false.[42] He points out that since overall similarity among worlds is *vague*, "some questions about how things would be if A have no non-arbitrary answers: if you had a sister would she like blintzes?"[43] Moreover, even without any vagueness it can be genuinely *indeterminate* which of two possible worlds is closer to actuality simply because the actual world is not determinate in all respects. For example, the "no-hidden variable" proofs in quantum mechanics (on some readings) show that there is no fact of the matter about whether a given photon will pass through a pane of glass or be reflected. The best one can do is to give the relevant probabilities. One is reduced to speaking in probabilistic terms here not just because one cannot *know* which of the two results will occur, but because *there is no fact to be known*. Now, imagine that we have two panes of glass set parallel to one another, as storm and interior windows are, and suppose that a certain photon is reflected off the first pane. It makes sense to consider the counterfactuals

[42] Lewis (1973, pp. 71–83).   [43] Lewis (1981, p. 23).

172

If the photon had passed through the first window pane, then it would have passed through the second.

If the photon had passed through the first window pane, then it would have been reflected back off the second.

Which is true? The only reasonable answer is that both are false because the facts of the actual world are not sufficiently determinate to settle the question of which possible photon-passes-through-the-first-pane world is closer to actuality.

I find Lewis's arguments persuasive, and so reject Conditional Excluded Middle. Gibbard and Harper recognize a problem here as well, and they write that in cases where Conditional Excluded Middle fails, "it would seem that using conditionals to compute expected utility is inappropriate. A more general approach is needed to handle such cases."[44] Lewis suggests that one part of the generalization is to replace each subjunctive conditional $A \mathbin{\square\!\!\rightarrow} W$ that appears in Stalnaker's formula by a family of conditionals of the form $A \mathbin{\square\!\!\rightarrow} CH(W) = x$, each of which specifies the chances that W would have of being true were A to be performed. The effect of this is to set $P(\bullet \backslash A) = \Sigma_x x P(A \mathbin{\square\!\!\rightarrow} CH(\bullet) = x)$, which makes A a more efficient cause of E than B is just in case the agent expects that E would have a higher unconditional chance if A were true than if B were true, and to replace the simple version of the Stalnaker formula by

$$\textbf{CDT}_{\text{SE-2}} \quad U(A) = \sum_W \left[\sum_x x P(A \mathbin{\square\!\!\rightarrow} CH(W) = x) \right] u(W)$$

This is surely the right way to handle the quantum-mechanics case, since exactly one chance conditional of the form

If the photon had passed through the first window pane, then its chance of passing through the second would be x.

will be true. Moreover, this way of rewriting Stalnaker's equation brings the Stalnaker/Gibbard/Harper version of causal decision theory into close contact with the "conditional chance" approach.[45]

---

[44] Gibbard and Harper (1978, pp. 344–45).

[45] Indeed, the theories might be identical when $P(A) > 0$. It is hard to see why anyone would want to distinguish a decision maker's expectation of E's chance conditional on A from her expectation of the unconditional chance E would have if A were true. It may not even make sense to do so. If the Principal Principle is right then no decision maker can consistently suppose that she might do A when the chance of E conditional on A differs from E's unconditional chance if A were true. For if $P(A$ & $CH(E/A) = x$ & $(A \rightarrow CH(E) = y)) > 0$ when $x \neq y$, then, since $(A$ & $(A \mathbin{\square\!\!\rightarrow} CH(E) = y))$ is equivalent to $(A$ & $CH(E) = y)$, PP seems to require that $P(E/A$ & $CH(E/A) = x$ & $(A \mathbin{\square\!\!\rightarrow} CH(E) = y))$ be equal to both x and y. Thus, it seems that an agent cannot coherently suppose that she will do A unless she also believes that $CH(E/A) = x$ & $(A \mathbin{\square\!\!\rightarrow} CH(E) = y))$ only holds when $x = y$.

We must realize, however, that Lewis's maneuver does not completely circumvent the problem of the Conditional Excluded Middle (nor does he claim it does). The principle can still fail either because the notion of overall similarity is inherent vague or because the actual facts are not sufficiently determinate to fix even a definite *subjunctive chance* for each event the decision maker might care about. Suppose that you were to have one more child than you actually ever do. What are the objective chances that she would grow up to be a dentist? Answer: There is no answer because the actual facts do not make any chance assignment reasonable. This indeterminacy presents a problem for CDT_{SE-2} because, in cases where an agent does not believe that some outcome W would have a definite chance were A true, the expectation $\Sigma_x xP(A \ \square\!\!\rightarrow CH(E) = x)$ will not be a well-defined magnitude, and A will lack an efficacy value.

Looking at all this indeterminacy, one may be tempted to draw the moral that causal utilities cannot be characterized subjunctively after all. This would be a mistake. The fact that counterfactuals, with or without chance consequents, sometimes violate Conditional Excluded Middle should not lead us to think *that subjunctive beliefs* must be similarly ill defined. We must be careful to distinguish a person's *subjective conditional probability for E on the subjunctive supposition that A* from both her unconditional subjective probability for the conditional $A \ \square\!\!\rightarrow E$, $P(A \ \square\!\!\rightarrow E)$, and her unconditional expectation for E's chance if A were true, $\Sigma_x xP(A \ \square\!\!\rightarrow CH(E) = x)$. When I suppose, counterfactually, that I will have another child I can perfectly well assign a low, but nonzero, probability to the proposition that she will be a dentist even though I assign the proposition "If I were to have a child, she would be a dentist" a probability of 0 (because I am sure that there is no world in which the antecedent is true that is "most similar to" the actual world), and even though I do not believe there is a definite objective probability for a possible child of mine becoming a dentist. The point here is that the epistemic operation of judging how likely E would be if A were true is different from that of judging how likely $A \ \square\!\!\rightarrow E$ is or that of estimating the objective chance that E would have if A were true. I will call the former operation *imaging on A* because, as we will see in the next chapter, it turns out to be a natural generalization of a belief revision method that Lewis has called by this name.[46] The imaging probability function will be denoted hereafter by P^A.

[46] Lewis (1976).

Imaging is the correct epistemic operation for calculating efficacy values. We get the correct value for $U(A)$ by identifying $P(\bullet \backslash A)$ with $P^A(\bullet)$ in CDT. When a decision maker evaluates A's efficacy value she provisionally modifies her opinions by supposing that A is true and redistributing her subjective probability for $\neg A$ over A-worlds in a way that respects her judgments about relevant overall similarity among worlds. We will talk about how this redistribution works in the next chapter, but the basic requirement is that worlds that are judged to be "less like the actual world" should get less additional weight than those "closer to actuality." Upon completing the epistemic operation of supposing A in this subjunctive way, the agent then computes A's utility relative to this new, provisional belief state to obtain its efficacy value. Thus, the right equation for calculating causal expected utilities is

$$\textbf{CDT} \quad U(A) = \sum_W P^A(W) u(W)$$

where P^A is the image of P on A.

The idea of expressing efficacy value in terms of imaging in this way is not new. J. Howard Sobel suggested it in the late 1970s, and the idea was discussed by Lewis as well.[47] I shall, however, be understanding the imaging function in a more general way than either of them does. As I propose to understand the notion, the core of the relationship between imaging and probabilities of subjunctive conditionals is given by

$$P(\neg(A \,\square\!\!\rightarrow \neg E)) \geq P^A(E) \geq P(A \,\square\!\!\rightarrow E)$$

The intuitive rationale for these inequalities is straightforward. One can think of $P(A \,\square\!\!\rightarrow E)$ as a measure of the agent's degree of confidence in E being *determinately true* if A were true. It is the probability she assigns to E when she supposes A *and imagines that some A-world is closer than any other to actuality*. $P(\neg(A \,\square\!\!\rightarrow \neg E))$, on the other hand, is her degree of confidence in E's *not being determinately false* if A were true. It is 1 minus the probability that she assigns to $\neg E$ when she supposes A and imagines that some A-world is closer to actuality than any other. The inequalities merely state that

---

[47] See Sobel (1994) and Lewis (1981). Lewis points out that this version of CDT reduces to a **K** partition view if dependency hypotheses are disjunctions of worlds with equal P^A-probability. However, this requires Stalnaker's Assumption and it applies only if $P(A) > 0$. Moreover, this reading seems to depart from the idea that dependency hypotheses are maximally specific causal stories. The fact that $P^A(W) = P^A(W^*)$ need not indicate that A has the same causal powers in W and W^*.

a rational agent cannot think that E is less likely to be true (false) simpliciter than it is to be determinately true (false) were A true. Since $P^A(E) = P(A \;\square\!\!\rightarrow E)$ when a closest A-world exists, Stalnaker's Equation will yield correct values for the agent's estimates of efficacy right provided that she believes that there is a unique closest A-world.

A similar pair of inequalities holds for estimates of objective chances:

$$1 - \sum_x xP(A \;\square\!\!\rightarrow \mathrm{CH}(\neg E) = x) \geq P_i^A(E)$$
$$\geq \sum_x xP(A \;\square\!\!\rightarrow \mathrm{CH}(E) = x)$$

Here, the expression on the right gives E's *determinate expected chance of being true* if A were true, and the one on the left gives its *determinate expected chance of not being false* if A were true. Insofar as one thinks that estimates of objective chance should inform rational opinion, one will see these as reasonable upper and lower bounds on $P^A(E)$. Again, when determinacy reigns, so that $1 - \Sigma_x xP(A \;\square\!\!\rightarrow \mathrm{CH}(\neg E) = x) = \Sigma_x xP(A \;\square\!\!\rightarrow \mathrm{CH}(E) = x)$, one will get the correct result that $P^A(E) = \Sigma_x xP(A \;\square\!\!\rightarrow \mathrm{CH}(E) = x)$.

For now this will have to do as a discussion of imaging. We will return to the topic in the next chapter. First, however, we need to address a more pressing issue having to do with any expected utility function that has the form CDT.

5.5 A GENERAL VERSION OF CAUSAL DECISION THEORY

The major shortcoming of CDT is that it only tells us how to calculate the utilities of actions relative to the "grand-world" partition of the atomic propositions of Ω. A **K**-partition formulation does not do any better since it too uses a special partition of propositions to compute efficacy values. As we saw in our discussion of Savage, it is unwise to formulate decision theory in this way because it makes it impossible to apply it to the kinds of "small-world" decision problems people face in real life. The only way to obtain a theory with such "small-world" applicability is by finding a partition-invariant equation for computing expected utilities. None of the versions of causal decision theory we have considered thus far is formulated in a partition-invariant way.

Some causal decision theorists are willing to swallow this limitation. David Lewis, for example, has written:

We ought to undo a seeming advance in the development of decision theory. Everyone agrees that it would be ridiculous to maximize the "expected utility" defined by $\Sigma_K P(Z)V(A \,\&\, Z)$ where Z ranges over just any old partition. It would lead to different answers for different partitions. . . . What to do? Savage suggested, in effect, that we make the calculation with unconditional [subjective probabilities], but make sure to choose only the right sort of partition. But what sort is that? Jeffrey responded that we would do better to make the calculation with conditional [subjective probabilities]. . . . Then we need not be selective about partitions, since we get the same answer, namely $V(A)$, for all of them. In a way, Jeffrey himself was making decision theory causal. But he did it by using [evidential] dependence as a mark of causal dependence, and unfortunately the two need not always go together. So I have thought it better to return to unconditional [subjective probabilities] and say what sort of partition is right.[48]

J. Howard Sobel goes even further, arguing that this "seeming advantage" of evidential decision theory is actually a disadvantage.

In Jeffrey's logic of decision there is no problem with adequate partitions of circumstances [our states and events]. In that theory, every partition of circumstances is adequate to the analysis of every action; this is one of the things that makes the theory attractive. Nobody likes problems. But this source of attractiveness is at the same time an embarrassment, for it is not really credible that every partition should be adequate. Jeffrey's theory lacks a problem that one expects a decision theory to have. Our causal decision theory does have a problem here, which is thus not only bad for it but good.[49]

Sobel, much to his credit, has gone on to prove some important "partition theorems" for causal decision theory, with the goal of characterizing the entire range of decision problems to which it can be reasonably applied.

Despite their pronouncements, Lewis and Sobel are wrong. The partition invariance of evidential decision theory is a real advantage (no "seeming" about it), and it in no way detracts from Jeffrey's account that it can be formulated in a partition-invariant way.[50] Indeed, as was suggested in previous chapters, a theory of rational choice that cannot be given a partition-invariant expression will be of no use to us in the decisions we face because it provides us with no clear criterion for calling "small-world" choices correct or incorrect. This is not, of course, to say that the causal theory is incorrect as it stands, only that it is incomplete in an important respect.

---

[48] Lewis (1981, p. 11), with some minor modifications in notation.

[49] Sobel (1994, p. 161).

[50] Ellery Eells makes this point in his (1985, p. 102). Isaac Levi concurs in his (1982, p. 342).

I suspect that Lewis and Sobel are willing to accept the lack of partition invariance because they think that no partition-invariant formulation of causal decision theory is in the offing. Fortunately, this is not so; there is a way of expressing causal decision theory that makes U values independent of the partition relative to which they are calculated. It only requires a simple mathematical trick to reformulate the theory in a partition-invariant way. My hunch is that causal decision theorists have not seen it mainly because doing so would force them to make what seems a major concession by granting that Jeffrey was right about how we desire things; it requires one to concede that *all value is news-value.*

For a causal utility to be formulated in a partition-invariant manner it must be possible to assign efficacy values in such a way that for any partition of Ω propositions $\{X_1, X_2, X_3, \ldots\}$ and any action A, one will have $U(A) = \Sigma_W P(W\backslash A)u(W) = \Sigma_j P(X_j\backslash A)U(A \ \& \ X_j)$. The only nonaccidental way for this to happen is by having $P(X_j\backslash A)U(A \ \& \ X_j) = \Sigma_W P(W \ \& \ X_j\backslash A)u(W)$ for every X_j. Since the X_js may be any propositions whatever, it follows that in a fully partition-invariant version of casual decision theory it will always be true that

$$U(A \ \& \ X) = \sum_W [P(W \ \& \ X\backslash A)/P(X\backslash A)]u(W)$$

Since $P(\bullet\backslash A) = P(A \ \& \ \bullet\backslash A)$, this expression is identical to Jeffrey's $V(A \ \& \ X)$ except that the decision maker's causal probability $P(\bullet\backslash A)$ has been substituted for her unconditional probability P. What one has here, then, is the *auspiciousness* of A & X computed relative to a causal probability. I propose to express causal decision theory in terms of "causalized" news values of this sort.

Definition. *A* general causal expected utility *for a decision problem* **D** *= (Ω, **O**, **S**, **A**) is a two-place, real-valued function*

$$V(X\backslash Y) = \sum_W [P(W \ \& \ X\backslash Y)/P(X\backslash Y)]u(W)$$

defined on (at least) $\Omega \times \Omega(\textbf{A})$ where u is an assignment of utilities to the atoms of Ω, and $P(X\backslash Y)$ is a casual probability function.

I have written this using a V rather than U to indicate that general causal expected utility is a kind of news value for each fixed Y. This being the case, $V(\bullet\backslash Y)$ will always be partition-invariant; that is, for any partition $\{X_1, X_2, X_3, \ldots\}$ of X one will have

$$V(X\backslash Y) = \sum_j [P(X_j\backslash Y)/P(X\backslash Y)]V(X_j\backslash Y)$$

178

What makes this partition invariance possible is the recognition that the process by which an agent determines the efficacy value of an act has two distinct stages. The first is an epistemic operation of *causal supposition* that involves provisionally adopting a new system of beliefs that appropriately expresses one's judgments about what the act would promote or inhibit. (These causal beliefs are the appropriate causal analogue of Price's "agent probabilities.") The second step involves a separate, *axiological* operation that evaluates the desirabilities of *all* the propositions in Ω in terms of their news values computed by using this new probability. Standard versions of causal decision theory recognize the first phase of this process but have too limited an understanding of the second. By focusing only on the casual expected utility of the act itself and ignoring the other propositions in Ω, causal decision theorists needlessly deprive themselves of the means of obtaining a partition-invariant formulation of their theory. One is not going to get the whole story about the good and bad things that A will cause by looking only at $U(A) = V(A\backslash A)$. One also needs to refer to auspiciousness values of the form $V(X\backslash A)$ where X might be *any* proposition consistent with A.

My way of formulating causal decision theory has a number of advantages aside from partition invariance. First, it helps to make clear what a decision maker is up to when she evaluates the causal expected utility of her acts. Notice that $U(A)$ can be written either as $V(A\backslash A)$ or as $V(\mathbf{T}\backslash A)$ where $\mathbf{T}$ is the necessary proposition. Recall that $\mathbf{T}$'s news value is the decision maker's subjective estimate of the overall desirability of the *status quo* (since any proposition whose truth is better than its falsity will have a news value above $\mathbf{T}$'s). This makes $V(\mathbf{T}\backslash A)$ a measure of the desirability of *the status quo that A would bring about*. Causal decision theory is the thesis that a person should try to make choices that maximize this value, that is, one should act in the way that one expects will leave one better off for having so acted. This, of course, is exactly right. Talking about A's efficacy value is really a way of talking about the (nonefficacy) value of the total state of affairs that A will bring about. One of the things I like most about writing the efficacy value of actions as causalized news values is that it makes this point crystal clear.

Another advantage is that it allows us to see evidential and causal decision theories as having a common underlying form. Both say that the value of an action A is found by looking at its news value as computed relative to some probability Q for which $Q(A) = 1$. The distinction between the two theories does not, therefore, have to do with any deep difference in their basic account of *valuing*. Rather, it concerns the appropriate *epistemic state* from which value judgments are to be made. Evidential decision theorists tell us to evaluate acts by provisionally modifying our opinions in the way that best reflects our considered judgments about their evidential import, whereas causal decision theorists tell us to make this modification in a way that best reflects our judgments about their causal powers. This is a significant difference, of course, but it is one that turns out to have rather little impact on the *formalism* of decision theory.

This is fortunate. A long-standing problem for causal decision theory has been its lack of an entirely acceptable representation result. In the next two chapters we will see that, once we understand causal decision theory in the way advocated here, it becomes possible to prove a version of Bolker's representation theorem to serve as its foundation. This provides, for the first time, an acceptable theoretical underpinning for our best account of decision making.

6

A General Theory of Conditional Beliefs

If the suggestion made at the end of the previous chapter is right, then the main differences between causal and evidential decision theory have nothing essential to do with their underlying theories of value; all value is news value. The disparity between the two positions has to do, rather, with a difference of opinion about the correct *epistemic perspective* from which rational agents should evaluate their options. Evidential decision theorists believe that actions should be assessed as if they were ordinary pieces of information about the world, whereas causal decision theorists maintain that they must be viewed as potential causes of desirable or undesirable outcomes. To make either of these assessments, an agent must engage in an epistemic operation called *supposition*. Evidential and causal decision theorists both maintain that actions should always be evaluated *under the supposition that they are performed*, but they have rather different notions of supposition in mind when they say this.

Proponents of the evidential theory are speaking about supposition in the *indicative* or *matter-of-fact mode*. This process involves provisionally accepting some proposition as true in a way that preserves its evidential relationships. The hallmark of indicative supposition is that it never forces the supposer to revise any of her views about the "facts," that is, propositions that she is certain are true. Causal decision theorists, in contrast, believe that people should evaluate their actions by *subjunctively* supposing that they will be performed. Here the person is guided not by her antecedent views about "the facts" as they actually happen to be, but by her judgments of similarity and difference among merely possible situations. The contrast between these two modes of supposition is just the difference between asking, "What is the world like if I do perform A?" and asking, "What would the world be like were I to perform A?"

The purpose of the present chapter is to clarify the ways in which these two modes of supposition differ and are alike. This investigation is preparatory to the next chapter, in which a general representation theorem that encompasses both causal and evidential decision theory is proved. Since this theorem will rely crucially on the abstract notion of a *conditional* belief, a notion that includes beliefs under both matter-of-fact and subjunctive suppositions, one goal of this chapter will be to determine the properties that all suppositions share. Another will be to draw the line between matter-of-fact and subjunctive supposition as clearly as possible, so as to highlight the core of the disagreement between the evidentialist and causalist approaches to decision theory.

6.1 THE CONCEPT OF A SUPPOSITION

As Bas van Fraassen has noted, we cannot adequately account for the full range of people's doxastic attitudes merely by speaking in terms of unconditional beliefs; we must also mention *suppositions*:

> Much of our opinion can only be elicited by asking us to suppose something, which we may or may not believe. The respondent imaginatively puts himself in the position of someone for whom the supposition has some privileged epistemic status. But if his answer is to express his present opinion ... this "momentary" shift in status must be guided by what his present opinion is.[1]

In general, a *supposition* is a form of provisional belief revision in which a person accepts some proposition C as true and makes the minimum changes in her other opinions needed to accommodate this modification.[2] Someone who supposes C acts *as if* she believes it for a time so as to see what its truth would involve when viewed from a special perspective.

There are, as we will see, a variety of senses in which belief revisions can be minimal, and thus a variety of ways in which propositions can be supposed. For our purposes the two most important of these ways are *indicative*, or *matter-of-fact*, supposition, and supposition in the *subjunctive* mode. Consider Ernest Adams's famous example of a man who thinks that Kennedy would *not* have been shot if Oswald had not shot him, but who also knows that Kennedy was shot even if Oswald did not shoot him.[3] Or, following Jonathan Bennett,[4] one might ask: Was *Hamlet* written if Shakespeare did not write it? Would

---

[1] van Fraassen (1995, p. 351).
[2] This is *provisional* belief revision. Someone who supposes C does not give up her current opinions. Her beliefs under the supposition of C and her genuine beliefs exist side by side, and she can go back and forth between the two.
[3] Adams (1965).
[4] Bennett (1988, pp. 523–24).

Hamlet have been written if Shakespeare had not written it? Unless you are one of those who think that Christopher Marlowe or the earl of Oxford wrote the plays commonly attributed to Shakespeare, you will answer "yes" to the first question and "no" to the second. Here you are supposing *in two distinct ways* that Shakespeare did not write *Hamlet*. When you suppose it *indicatively* you ask how things *are* if, as a matter of fact, Shakespeare did not write the play, and, since you know the play was written by someone, you conclude that it must have been someone other than Shakespeare. On the other hand, when you suppose that Shakespeare did not write *Hamlet* in the *subjunctive* mode you ask how things *would have been* in a counterfactual situation in which Shakespeare did not write *Hamlet*, and, as Shakespeare was a singular genius, you conclude that the play would not have been written at all. As we will soon see, these two species of supposition involve preserving certain aspects of one's beliefs about conjunctions and subjunctive conditionals, respectively. They are represented probabilistically by standard Bayesian conditioning and by imaging.

Throughout our discussion I will be employing a probabilist framework that assumes agents with graded or "partial" beliefs that meet the minimum Bayesian requirement of rationality: probabilistic representability. This is somewhat out of step with the current literature. Some of the best recent work on belief revision has been done within the confines of *dogmatic*[5] epistemology, which sees *full belief*, the categorical acceptance of some proposition as true, as the fundamental doxastic attitude.[6] We will be making some contact with this literature in the course of our discussions of certainties and null propositions, but, since the full story of our doxastic lives involves states of partial belief as well as states of acceptance, it behooves us to take a more general approach to the problem of belief revision. Belief revision will therefore be portrayed here as a process by which a subject moves from one system of partial beliefs, modeled by a comparative likelihood ranking or a subjective probability, to another set of partial beliefs characterized in the same way. It will be useful to deal with the quantitative case first since this will supply us with the concept of a representation needed in the analysis of suppositions that are characterized qualitatively.

While our primary concern will be with cases in which a person supposes a single proposition to be true, it will be useful to start from

---

[5] I am taking this term from Jeffrey (1991).
[6] See, for example, Gardenfors (1988).

a more abstract perspective that treats suppositions as instances of *constrained belief revision.*[7] This is an epistemic process, much discussed in the literature, in which a person with subjective probability P provisionally modifies her opinions by first imposing a *constraint* that fixes important features of her final belief state and then selecting a new probability from among those that satisfy the constraint by minimizing some function that measures "divergence" from P. In this context a *constraint* is any *convex, closed* subset Φ of the set π of all probability functions defined on P's base algebra Ω.[8] Such constraints can always be expressed in terms of equalities and inequalities that fix expected values of random variables computed using the new probability. All have the form $\Phi = \{Q \in \pi: \text{Exp}_Q(g_1) \geq x_1 \ \& \ \dots \ \& \ \text{Exp}_Q(g_n) \geq x_n\}$ where the g_j are (measurable) real functions on Ω, the x_j are real numbers, and $\text{Exp}_Q(g_j)$ is g_j's expected value computed using Q. Here are some examples of constraints that can be specified in this way:

- Some proposition C is true: $\Phi = \{Q \in \pi: Q(C) = 1\}$
- Some proposition C has probability q: $\Phi = \{Q \in \pi: Q(C) = q\}$
- Probabilities are distributed over some countable partition $\{C_1, C_2, C_3, \dots\}$ in a particular way:

$$\Phi = \{Q \in \pi: Q(C_1) = q_1 \ \& \ Q(C_2) = q_2 \ \& \dots \& \ Q(C_n) = q_n\}$$

- The conditional probability of X given C is q:

$$\Phi = \{Q \in \pi: Q(X/C) = q\}$$
$$= \{Q \in \pi: Q(X \ \& \ C)(1 - q) - Q(\neg X \ \& \ C)q = 0\}$$

In this highly abstract context, belief revision involves the application of a *selection rule* that assigns a revised final probability $P(\bullet \| \Phi)$ to each pair consisting of a prior probability P and a constraint Φ. This selection rule is required to satisfy the following conditions:[9]

- $P(\bullet \| \Phi) \in \Phi$.
- If $P \in \Phi$, $P(\bullet \| \Phi) = P(\bullet)$.
- If $P(\bullet \| \Phi) \in \Phi^*$ and $P(\bullet \| \Phi^*) \in \Phi$, then $P(\bullet \| \Phi) = P(\bullet \| \Phi^*)$.

[7] I am taking this term from Diaconis and Zabell (1983). A particularly useful treatment of belief revision in a probabilistic context may be found in Skyrms (1987).

[8] That is, it is a subset of π that contains the line segment between any two of its points, so that $P_1, P_2 \in \pi$ imply $\lambda P_1 + (1 - \lambda)P_2 \in \pi$ for all $\lambda \in [0, 1]$, and which contains its limit points.

[9] This definition comes from Skyrms (1987, p. 228).

The first of these is just the obvious requirement that the selected function $P(\bullet\|\Phi)$ should satisfy the constraint in question. The second expresses the thought that P always diverges minimally from itself. The last requirement rules out the possibility of there being two probabilities such that the first diverges less from P relative to one constraint that they both satisfy while the second diverges less from P relative to another such constraint.

When we focus on a constraint that forces some proposition C to be taken as certain, so that $\Phi = \{Q \in \pi: Q(C) = 1\}$, we can write these conditions as follows (using the notational convention that $P(\bullet\|C) = P(\bullet\|\{Q \in \pi: Q(C) = 1\})$):

- $P(C\|C) = 1$.
- If $P(C) = 1$, then $P(\bullet\|C) = P(\bullet)$.
- If $P(C\|C^*) = 1$ and $P(C^*\|C) = 1$, then $P(\bullet\|C) = P(\bullet\|C^*)$.

All of these facts will be familiar since they all remain valid when the double-bar "$\|$" is replaced by the backslash "/" that indicates an ordinary conditional probability. This is no accident for, as we shall soon see, ordinary conditional probabilities can be understood in terms of selection rules.

One generally thinks of each selection rule as determined by a *similarity gauge*. This is a mapping $\sigma(P, Q)$ that takes pairs of probabilities to non-negative real numbers which measure the degree to which the values assigned by Q *diverge* from those assigned by P. The inequality $\sigma(P, Q) > \sigma(P, Q^*)$ signifies that P is "more like" Q than Q*. Any reasonable similarity gauge must have at least the following two properties:

Centering. $\sigma(P, Q) = 0$ if and only if $P = Q$.
Convexity. $\sigma(P, \lambda Q_1 + (1 - \lambda)Q_2) \leqslant \lambda\sigma(P, Q_1) + (1 - \lambda)\sigma(P, Q_2)$ for all $\lambda \in [0, 1]$.

Centering says that no probability is more like itself than itself. To appreciate what Convexity means, suppose that, according to σ, Q_2 diverges from P more than Q_1 does, and consider their *even mixture* $\frac{1}{2}Q_1 + \frac{1}{2}Q_2$, which assigns each proposition a value half-way between those assigned by Q_1 and Q_2. Assuming that similarity gauges are continuous,[10] Convexity is equivalent to the statement that $\frac{1}{2}Q_1 + \frac{1}{2}Q_2$ is more like P than Q_2 is. To see how odd it would be for this to fail,

[10] This is actually a consequence of Convexity.

185

notice that to move from Q_2 to $\frac{1}{2}Q_1 + \frac{1}{2}Q_2$ one must change each $Q_2(X)$ value by an increment of $\frac{1}{2}Q_2(X) - \frac{1}{2}Q_1(X)$, and one must do the same thing to move from $\frac{1}{2}Q_1 + \frac{1}{2}Q_2$ to Q_1. So, if $\sigma(P, \bullet)$ were not convex one could start from Q_1 and change one's probabilities in a way that makes them less like P, and then make an *identical* shift and end up closer to P.

Convexity guarantees that $\sigma(P, \bullet)$ has a *unique* minimum on each convex constraint set Φ in π. This lets us associate a selection rule with σ by designating $P(\bullet\|\Phi)$ as the unique probability $Q \in \Phi$ at which $\sigma(P, \bullet)$ assumes its minimum, so that $\sigma(P, Q) \geq \sigma(P, P(\bullet\|\Phi))$ for all $Q \in \Phi$. It follows from Centering that $P(\bullet\|\Phi)$ and $P(\bullet)$ must coincide when $P \in \Phi$. And, since $P(\bullet\|\Phi)$ minimizes $\sigma(P, \bullet)$ subject to Φ it follows directly that $P(\bullet\|\Phi)$ and $P(\bullet\|\Phi^*)$ must be identical when $P(\bullet\|\Phi) \in \Phi^*$ and $P(\bullet\|\Phi^*) \in \Phi$. Thus, the function $P(\bullet\|\bullet)$ will always be a selection rule when σ is a similarity gauge.

All this talk of similarity gauges and selection rules should remind one of Stalnaker's semantics for subjunctive conditionals. Indeed, as Brian Skyrms has observed,[11] it *is* the Stalnaker semantics with probabilities serving as surrogates for possible worlds, constraints playing the role of propositions, $\sigma(P, Q)$ standing in for the similarity relation, and $P(\bullet\|\Phi)$ being the "closest" world in which Φ is true. Stalnaker's assumption, which requires the existence of a unique closest possible world in which some proposition obtains, appears now in the guise of the convexity condition. If we relax this assumption by allowing nonconvex similarity gauges, then we obtain a Lewis-style semantics in which there is not always a unique closest "world" in which Φ holds.

There is, in my view, one further property that similarity gauges should possess. It captures the idea that "divergence" among probability functions is solely a matter of differences in the values they assign. It seems clear that P should be classified as more like Q_1 than Q_2 when P's values are *uniformly closer* to Q_1's values than to Q_2's values. Indeed, something slightly stronger is true. Say that P is *uniformly closer to* Q_1 *than to* Q_2 *relative to a partition* Z when either

$$Q_2(X \& Z) \geq Q_1(X \& Z) \geq P(X \& Z)$$

or

$$P(X \& Z) \geq Q_1(X \& Z) \geq Q_2(X \& Z)$$

[11] Skyrms (1987, pp. 228–29).

holds for all $X \in \Omega$ all $Z \in \mathbf{Z}$. In other words, P is uniformly closer to Q_1 to than to Q_2 relative to $\mathbf{Z}$ when, for each cell Z of $\mathbf{Z}$, the probabilities P assigns to propositions that entail Z are always closer to their Q_1 probabilities than their Q_2 probabilities.[12] The condition I want to impose on similarity gauges is this:

Weak Uniform Dominance.[13] If P is uniformly closer to Q_1 than to Q_2 relative to any partition $\mathbf{Z}$, then $\sigma(P, Q_2) \geq \sigma(P, Q_1)$.

To deny this is to reject the idea that similarity among probability functions is a matter of similarities among the values they assign. It is hard to see what such a notion of similarity would be if it lacked this feature. It certainly would not be one capable of capturing the notion of supposition, which is our primary interest here. Henceforth we will assume that every similarity gauge satisfies this requirement.

In the context of the abstract model we have been developing, a supposition is a belief revision subject to a constraint of the form $\Phi = \{Q \in \pi: Q(C) = 1\}$. Since not every such constraint makes sense for every type of belief revision (e.g., $\{Q \in \pi: Q(C \ \& \ \neg C) = 1\}$ never does), we will usually stipulate that the supposed proposition C must lie in some distinguished subset $\mathbf{C}$ of *conditions* in Ω whose elements are apt for supposition. While different sets of conditions make sense for different notions of supposition, it is mathematically convenient to require the complement of $\mathbf{C}$ to be an *ideal*, so that $\Omega \sim \mathbf{C}$ contains the contradictory proposition $X \ \& \ \neg X$ and is closed under both countable disjunction and logical strengthening.[14]

Relative to such a family of conditions $\mathbf{C}$, the *supposition function* for P induced by the gauge σ is the function $P(\bullet \| \bullet)$ on $\Omega \times \mathbf{C}$ defined by $P(\bullet \| C) = P(\bullet \| \{Q \in \pi: Q(C) = 1\})$. The basic properties of suppositions are given by the following result:

Theorem 6.1. *If σ is a similarity gauge, then its associated supposition function obeys*

---

[12] Readers are invited to convince themselves that if P is uniformly closer to Q_1 than to Q_2 relative to $\mathbf{Z}$ then it will also be uniformly closer to Q_1 relative to any finer partition (but *not* necessarily relative to a coarser one). So, for an algebra that contains a countable number of atoms the requirement is that $Q_1(W)$ should fall between $P(W)$ and $Q_2(W)$ for every atom W.

[13] This makes sense because it will never happen that P is uniformly closer to Q_1 on every element of one partition and uniformly closer to Q_2 on every element of another.

[14] Closure under logical strengthening requires $X \ \& \ Y$ to be in $\Omega \sim \mathbf{C}$ whenever $X \in \Omega \sim \mathbf{C}$ and $Y \in \Omega$.

SUP₁ *(Coherence): P(•‖C) is a countably additive probability on Ω.*
SUP₂ *(Certainty): P(C‖C) = 1.*
SUP₃ *(Regularity): P(X‖C) ≥ P(C & X) as long as C ∈ **C**.*

The proof of this result is rather involved. It can be found in the Appendix at the end of this chapter.

SUP₁–SUP₃ are all properties that any reasonable notion of supposition ought to have. Insofar as we think that probabilistic coherence is an important constraint on rational belief, we should want to impose it on suppositional beliefs as well as unconditional ones since the laws of reason remain fixed whether one supposes that some proposition is true or not.[15] This is precisely what SUP₁ does. SUP₂ merely says that a person supposes C by provisionally adopting an attitude in which she regards C as certainly true. This distinguishes supposition from forms of provisional belief that do not involve accepting any proposition as true. SUP₃ is a more substantive condition. It says that if you assign a given probability to X & C before you suppose C, then you cannot assign a higher probability to X after you suppose C. This rules out absurdities like the following: The Dow Jones average is unlikely to fall if there is a recession, but there is likely to be a recession during which the Dow Jones average falls.

SUP₃ has a number of important consequences. First, it shows that P(•‖•) is "anchored" on the unconditional probability P(•) in the sense that P(•‖T) = P(•). This, of course, is exactly how it should be since supposing that a tautology is true should never alter one's beliefs. Even more important is the fact that SUP₃ safeguards the use of *conditional proof* within the scope of a supposition. Here is an equivalent way of stating the principle:

SUP₃ *("Conditional Proof" Version): P(C ⊃ X) ≥ P(X‖C) when*
    *C ∈ **C**.*

This says that if you assign X a given subjective probability on the supposition that C, then you should assign at least as high an unconditional probability to the material conditional C ⊃ X. This underwrites a common form of reasoning in which we justify a conditional proposition by arguing that its consequent is likely on the assumption that its antecedent obtains. We will discuss this sort of inference further in the next section.

[15] This is one way in which supposition differs from what Ken Walton calls "make believe," which need not respect the rules of rationality. One can, for example, make believe that one is looking at something that is black and white all over, but it is incoherent to suppose such a thing. See Walton (1990).

SUP_1–SUP_3 may be the only completely general properties shared by all forms of quantitative supposition.[16] To transport them into a comparative setting we need to imagine that a person's beliefs under the scope of a supposition can be described by a *conditional* comparative likelihood ranking $(\|.>.\|, \|.\geq.\|)$ whose basic terms are

$X\|C .>. Y\|D =_{df}$ The agent is more confident of X when she supposes that C than she is of Y when she supposes that D.

$X\|C .\geq. Y\|D =_{df}$ The agent is at least of confident of X when she supposes that C as she is of Y when she supposes that D.

As in the quantitative case, we require all propositions that appear as conditions in $(\|.>.\|, \|.\geq.\|)$ to be found in a subset $\boldsymbol{C}$ of Ω whose complement is a σ-ideal. The analogues of SUP_1–SUP_3 are

$\text{SUP}_1\text{*}$: For each condition C in $\boldsymbol{C}$, the unconditional likelihood ranking defined by

$$X .>^C. Y \text{ iff } X\|C .>. Y\|C$$

$$X .\geq^C. Y \text{ iff } X\|C .\geq. Y\|C$$

(hereafter to be called the *C-section* of $(\|.>.\|, \|.\geq.\|)$) must conform to all the laws of rationality to which any unconditional likelihood ranking must conform. In particular, it should be probabilistically representable.

$\text{SUP}_2\text{*}$: $C\|C .=. D\|D$ and $\neg C\|C .=. \neg D\|D$ for any $C, D \in \boldsymbol{C}$.
$\text{SUP}_3\text{*}$: $(C \supset X)\|\mathbf{T} .\geq. X\|C .\geq. (C \& X)\|\mathbf{T}$ for all $C \in \boldsymbol{C}$ and $X \in \Omega$.

Each of these conditions is necessary if $(\|.>.\|, \|.\geq.\|)$ is going to be represented by a probabilistic supposition function. Together they provide a partial specification of the laws governing supposition in the comparative mode. We will add to them as we go on.

6.2 SUPPOSITION AND CONDITIONALS

As noted, the regularity principle SUP_3 underwrites a common form of reasoning in which a person justifies a conditional proposition by arguing that its consequent is likely to be true on the supposition that its antecedent is true. It is, however, an extremely weak constraint on such reasoning, far weaker than a more famous requirement called the *Ramsey Test*. Frank Ramsey once proposed that the right degree

[16] This is not quite true. SUP_3 generalizes to the requirement that $P(X\|C \& C\text{*})$ should fall between $P(C \supset X\|C\text{*})$ and $P(X \& C\|C\text{*})$ whenever $(C \& C\text{*}) \in \boldsymbol{C}$.

of belief for a person to assign to a conditional is the one that she would obtain by adding its antecedent to her stock of knowledge, making the minimal revisions in her other beliefs needed to preserve consistency, and making a judgment about the likelihood of the consequent from that standpoint.[17] Some people who think they have an independent handle on the notion of minimal belief revision have tried to use this test as a way of providing a semantics for certain kinds of conditionals (e.g., indicative conditionals). Others, who think that they understand certain kinds of conditionals (e.g., subjunctive), have used Ramsey's test as a way of cashing out the concept of a minimal revision of belief.

In the current, probabilistic context we can express Ramsey's idea as follows:[18]

Ramsey Test. Any legitimate form of supposition can be associated with a conditional operator $\Rightarrow$ whose semantics ensure that $P(C \Rightarrow X) = P(X \| C)$ for all probability functions P defined on Ω and all conditions $C \in \boldsymbol{C}$.

If one thinks of a supposition function as a kind of conditional probability, this says that conditional probabilities are always analyzable as unconditional probabilities of conditional statements. SUP_3 is a much weaker requirement than Ramsey Test since it merely imposes upper and lower bounds on the probabilities of Ramsey conditionals *were they to exist*. As noted in Chapter 2, all conditionals are intermediate in strength between conjunction and the material conditional, so that $C \& X$ entails $C \Rightarrow X$ and $C \Rightarrow X$ entails $C \supset X$. Thus, a minimum condition for the Ramsey Test to hold is that it must always be the case that $P(C \supset X) \geq P(X \| C) \geq P(X \& C)$, which is precisely what SUP_3 requires.

It is fortunate that SUP_3 is weaker than the Ramsey Test because the latter is false for most important forms of supposition. Philosophers have long been trying to understand the semantics of the indicative or matter-of-fact conditional "If C is the case then X is the case." The first big breakthrough came with Ernest Adams's observation that people tend to *assert* an indicative conditional just in case their probability for its consequent conditional on its antecedent is high. In the ordinary case, of course, an assertion is meant to convey

---

[17] Excellent discussions of the Ramsey Test may be found in Gardenfors (1986) and Stalnaker (1984).

[18] This version of the Ramsey Test is probably stronger than anyone has ever proposed. Most discussions of the test take place within the context of dogmatic epistemology, and they tend to be restricted to discussions of indicative conditionals.

the speaker's belief that some proposition is true. This makes it quite natural to treat Adams as having shown that people who utter indicative conditionals are trying to assert *propositions* (though this is something Adams denied from the start), and that these assertions are governed by what Alan Hajek and Ned Hall call the *conditional construal of conditional probability*:

CCCP. There exists an indicative conditional $\rightarrow$ such that for any probability function P one has $P(C \rightarrow X) = P(X/C) = P(X \& C)/P(C)$ for any two propositions X and C such that $P(C) > 0$.

This is a version of Ramsey Test. As we will see, indicative supposition involves moving to a probability that agrees with $P(X/C)$ when $P(C) > 0$, and given this, CCCP amounts to the claim that there must be a Ramsey conditional for indicative supposition. It would be wonderful if this were true, for it would either allow us to explain the semantics for indicative conditionals by coming to understand how indicative supposition works or (what would be better for present purposes) allow us to characterize indicative supposition in terms of unconditional beliefs about indicative conditionals.

Unfortunately, there are few philosophical theses that have been more decisively refuted than CCCP. A classic argument due to David Lewis shows that no semantic operation can possibly fulfill $\rightarrow$'s job description.[19] Rather than recapitulating Lewis's reasoning it will be instructive to look at a slightly modified version of his argument that turns out to be quite revealing for present purposes. Since a supposition function $P(\bullet\|C)$ is always a probability it is a consequence of Ramsey Test that a Ramsey conditional will always satisfy the following two laws for any probability function P defined on Ω and any conditions $C \in \boldsymbol{C}$:

Probabilistic Conditional Contradiction. If X and X^* are contraries, then $P((C \Rightarrow X) \& (C \Rightarrow X^*)) = 0$.

Probabilistic Conditional Excluded Middle. $P((C \Rightarrow X) \vee (C \Rightarrow \neg X)) = 1$.

The core problem with CCCP is that no conditional for which $P(C \rightarrow X) = P(X/C)$ can possibly satisfy the latter of these two principles relative to a probability P for which both $P(X \& C)$ and $P(C \& \neg X)$ are nonzero and $P(C)$ is less than 1. To see why, note first that

---

[19] Lewis's argument has been elucidated and extended by Allan Gibbard, Peter Gardenfors, Alan Hajek, and Ned Hall, among others.

$$(*) \quad P(C \to X) = P(C \& X)P((C \to X)/(C \& X))$$
$$+ P(C \supset \neg X)P((C \to X)/(C \supset \neg X))$$

Employing Lewis's "key maneuver"[20] we can use CCCP to write

$$P((C \to X)/(C \& X)) = P(X/C \& (C \& X)) = 1$$
$$P((C \to X)/(C \supset \neg X)) = P(X/C \& (C \supset \neg X)) = 0$$

Hence, (*) reduces to $P(C \to X) = P(C \& X)$, and Conditional Excluded Middle then implies that $P(C \to \neg X) = P(C \supset \neg X)$. A symmetrical argument shows that $P(C \to X) = P(C \supset X)$ and $P(C \to \neg X) = P(C \& \neg X)$. Thus, for CCCP to hold it would have to be the case that

$$P(C \supset X) = P(C \to X) = P(C \& X)$$
$$P(C \supset \neg X) = P(C \to \neg X) = P(C \& \neg X).$$

But neither of these things can happen when $P(C) < 1$ since in such a case one will definitely have both $P(C \supset X) = P(\neg C) + P(C \& X) > P(C \& X)$ and $P(C \supset \neg X) = P(\neg C) + P(C \& \neg X) > P(C \& \neg X)$. Thus, there can be no Ramsey conditional for indicative supposition.

There is a diversity of opinion about what moral to draw from Lewis's argument. Some philosophers take it to show that speakers who utter indicative conditionals do not intend to assert propositions at all.[21] Indicative conditionals, it is said, lack truth conditions. Others maintain that, although asserters of indicative conditionals do express propositions, the proposition expressed varies with the beliefs of the person doing the asserting. Still others, notably Frank Jackson,[22] argue that indicative conditionals do have asserter-invariant truth conditions, but that these do not coincide with their assertability conditions. According to Jackson, the indicative conditional construction serves as a linguistic convention that is used when one wants to state that the corresponding material conditional is true, and to indicate that it is "robust" with respect to information about antecedent. Thus, in asserting, "If Shakespeare didn't write *Hamlet* some other Englishman did," I am conveying that I have a high degree of confidence in the disjunction "Shakespeare or some other Englishman wrote *Hamlet*" and implying that I will continue to have a high degree of confidence in disjunction even if I learn that the play was not written by Shakespeare. Since a person's belief in $C \supset X$ is robust

[20] I am taking this term from Hajek and Hall (1994).
[21] See, for example (Gibbard 1980).
[22] Jackson (1979).

under changes in her information about C just in case $P(C \supset X/C) = P(X/C)$ is high, this explains Adams's observation that the assertability of an indicative conditional is a function of the probability of its consequent conditional on its antecedent.

We need not spend time trying to adjudicate these issues. For us, the important thing about Lewis's result is not its impact on the semantics and pragmatics of conditionals, but its implications for our understanding of belief revision. The spin I would want to put on the matter is that CCCP goes wrong by supposing that a single semantic operation can do the jobs of both conjunction and material implication. Simple calculations show that conjunction and material implication provide a lower and an upper bound on conditional probabilities in the sense that the inequalities $P(C \supset X) \geq P(X/C) \geq P(C \& X)$ hold for any probability function P that assigns a nonzero probability to C. Moreover, as the following result shows, these bounds are the best that can be attained.

Theorem 6.2. *Except in the extreme situation where C logically entails X or ¬X (in which case no interesting conclusions can be drawn), if $ and @ are any sentential connectives such that the inequalities $P(C \$ X) \geq P(X/C) \geq P(C @ X)$ hold for all probabilities with $P(C) > 0$, then $C \supset X$ entails $C \$ X$ and $C @ X$ entails $C \& X$.*

(The proof may be found in the Appendix to this chapter.) Theorem 6.2 shows that a Ramsey conditional for conditional probability, for which $P(C \rightarrow X) \geq P(X/C)$ and $P(X/C) \geq P(C \rightarrow X)$ would have to hold whenever $P(C) > 0$, must be both logically stronger than conjunction and logically weaker than the material conditional. Of course, no such operation exists.

Theorem 6.2 also hints at an important general truth about the nature of supposition. Every legitimate form of supposition, I believe, can be associated with a characteristic kind of binary sentential operator, which, for want of a better term, I will refer to as its "conditioner." If @ is the relevant conditioner for a certain form of supposition, then a person's unconditional beliefs about statements of the form $C @ X$ will constrain her beliefs under the supposition that C is true. The rule is this:

Weak Ramsey Test. Any legitimate form of supposition can be associated with a conditioner @ whose semantics ensure that the inequalities $P(\neg(C @ \neg X)) \geq P(X\|C) \geq P(C @ X)$ hold for

193

all probability functions P defined on Ω and all conditions $C \in$ **C**. Moreover, $P(\neg(C @ \neg X))$ and $P(C @ X)$ serve as the best possible upper and lower bounds on $P(X\|C)$ in the sense that any other operation that satisfies the stated inequalities will be logically stronger than @.

In other words, for any form of supposition there is a "best possible" version of the Ramsey Test. Theorem 6.2 tells us that *conjunction* is the right conditioner for supposition in the indicative mood. In the next section we will see that the subjunctive conditional is the appropriate conditioner for subjunctive supposition.

While I have no general theory of conditioners to offer at the moment, I can say that all of them must share a number of logical properties with genuine conditionals. In the terminology of Chapter 2, any conditioner @ should satisfy at least

Centering. C & $(C @ X)$ is equivalent to C & X.

Weakening the Consequent. $C @ X$ entails $C @ (X \vee Y)$.

Conditional Conjunction. $(C @ X)$ & $(C @ Y)$ entails $C @ (X \& Y)$.

Conditional Contradiction. If X and X^* are contraries, then C & X and C & X^* are also contraries.

The first of these is required to ensure that the supposition function associated with @ will satisfy SUP$_3$, the principle that $P(X\|C) \geq P(C \& X)$. The last three are necessary for $P(\bullet\|C)$ to be a probability. Notice that Conditional Contradiction guarantees that $C @ X$ entails $\neg(C @ \neg X)$, and thus that it will always be possible for $P(X\|C)$ to fit beneath $P(\neg(C @ \neg X))$ and above $P(C @ X)$.

Once one understands the relationship between beliefs under a supposition and unconditional beliefs in the manner being suggested here it becomes clear that the failure of the Ramsey Test can be traced ultimately to the fact that conditioners do not generally satisfy the law of

Conditional Excluded Middle. $(C @ X) \vee (C @ \neg X)$ is a logical truth.

In cases where this law fails $C @ X$ and $\neg(C @ \neg X)$ will not be equivalent. Since the former entails the latter, and since a Ramsey conditional would entail the former and be entailed by the latter, it follows that no such conditional can exist. There is thus a tight linkage between the Ramsey Test and the law of Conditional Excluded Middle. The version of Lewis's impossibility proof that was presented here was chosen to highlight this connection.

There is, of course, much more that could be said about the rela-

194

tionship between conditional beliefs and conditional propositions, but pursuing these issues further would distract us from our main objective. Let us turn, therefore, to the task of characterizing the differences between indicative and subjunctive supposition.

6.3 INDICATIVE AND SUBJUNCTIVE SUPPOSITION CHARACTERIZED

As many of the preceding remarks have suggested, supposition in the matter-of-fact mode proceeds by simple conjunction. A person who supposes C simply conjoins its content to the content of her other beliefs, and then realigns her subjective probabilities in a way that best preserves her antecedent views about the likelihoods of these conjunctions. Her goal, in other words, is to move provisionally to a new probability function for which it is true that

MOF. If $P(C) > 0$, then $P(X \| C) \geq P(Y \| C)$ if and only if

$$P(C \& X) \geq P(C \& Y)$$

The rationale for this is straightforward: Insofar as one is treating C as a matter of fact, one wants to see what the world, as it is, will be like if C truth were added to the facts as they stand. This "matter-of-fact" character of MOF is revealed by following easily proved consequences:

Preservation of Certainties. If $P(X) = 1$ and $P(C) > 0$, then $P(X \| C) = 1$.

Symmetry. If $(D \& C) > 0$, then

$$P(X \& C \| D) \geq P(Y \& C \| D) \text{ iff } P(X \& D \| C) \geq P(Y \& D \| C)$$
$$\text{iff } P(X \| D \& C) \geq P(Y \| D \& C).$$

The first of these says that things taken as facts before a supposition is made (i.e., propositions assigned probability 1) are taken as facts afterward. The second principle is a comparative version of Bayes's theorem that explains how suppositions based on stronger conditions are related to suppositions based on weaker ones.

It turns out that MOF determines $P(\bullet \| C)$ *uniquely* when $P(C) > 0$. While there can be other belief revision rules consistent with MOF when P is defined on a small algebra of propositions, the ordinary conditional probability $P(\bullet/C)$ is the only rule that obeys MOF in cases where the set $\{X \in \Omega : P(X) > 0\}$ is atomless (so that each X with positive probability can be partitioned into disjoint propositions of

195

positive probability).[23] The only supposition functions *generally* capable of satisfying MOF must therefore be ones that coincide with the ordinary conditional probability when the proposition being supposed has positive probability.

Since MOF is itself a comparative condition it is easy to transform it into a constraint on likelihood rankings. We can characterize a person's beliefs when she supposes in the matter-of-fact mode in terms of an *indicative* conditional likelihood ranking $(//.>.//, //.\geq.//)$ that satisfies $SUP_1*–SUP_3*$ above as well as

MOF(comparative version). If $C//D.>.\neg T//D$, then $X//(C \& D)$

$.\geq.Y//(C \& D)$ if and only if $(C \& X)//D.\geq.(C \& Y)//D$.

If $(//.>.//, //.\geq.//)$ is going to be probabilistically representable, then MOF is necessary (and, in the case where the ranking is atomless, sufficient) for it to be represented by a supposition function $P(\bullet\|D)$ that assumes the form of a conditional probability for any condition C that is *nonnull* in the sense that $C//D.>.\neg T//D$.

Notice how MOF leaves the definition of $P(\bullet\|C)$ open when $P(C) = 0$. What I have said is that a probabilistic representation of matter-of-fact supposition must agree with $P(\bullet/C)$ *over the condition set* $\{C \in \Omega: P(C) > 0\}$, but I have not gone further to say that this set contains the *only* legitimate objects of matter-of-fact supposition. My view is that it can make sense to suppose propositions *as facts* even when one is certain that they are false. I admit that this sounds strange. Shouldn't matter-of-fact supposition only make sense for propositions the supposer regards as having some chance of actually *being* facts? Doesn't subjunctive supposition always apply when one is sure that the proposition being supposed is contrary to the facts? In the next section I will argue that this is not the case. Before I can do so, however, we need a better grip on the concept of subjunctive supposition.

The first step lies in getting clear about the operation of *imaging*, which was briefly mentioned in the previous chapter. Imaging was introduced by David Lewis to explain how probabilities are assigned to subjunctive conditionals.[24] Under the simplifying hypothesis that similarity among "possible worlds" (i.e., atomic elements of Ω) obeys Stalnaker's Assumption (and thus Conditional Excluded

---

[23] The proof, which will follow from Theorem 7.2 of the next chapter, is an application of the Ordinal Uniqueness Theorem of Chapter 4, and the fact that any $P(\bullet\|C)$ that obeys MOF will be ordinally similar to $P(\bullet/C)$. Paul Teller proved essentially this result in his (1973, pp. 227–32).

[24] Lewis (1976, pp. 308–15).

Middle), Lewis proved that $P(C \;\square\!\!\rightarrow X)$ will always coincide with the value of *the image of P on C*. This is the probability function defined by

Imaging. $P(X \backslash C) = P^C(X) = \sum_W P(W)P(X/W_C)$

where the sum is taken over all the "worlds" in Ω and W_C is the "world" (whose existence is guaranteed by Stalnaker's Assumption) that is most similar to W in which C obtains.[25] The best way to see what this means is by noticing that, for each world W^*, one will have $P^C(W^*_C) = \sum_W P(W)P(W^*_C/W_C)$. Since $P(W^*_C/W_C) = 1$ when $W_C = W^*_C$ and 0 otherwise, this tells us that imaging on C works by transferring the probability of each world in which C is false to the most similar world in which C is true. As Lewis points out, this makes $P(\bullet \backslash C)$ a kind of minimal revision of P, because "imaging P on C . . . involves no gratuitous movement of probability from worlds to dissimilar worlds."[26] To see why $P(X \backslash C)$ and $P(C \;\square\!\!\rightarrow X)$ must coincide when Stalnaker's Assumption holds note that, in this case, $C \;\square\!\!\rightarrow X$ will be the disjunction of worlds W in Ω such that W_C entails X. Since a world entails X exactly if $P(X/W_C) = 1$ and $P(\neg X/W_C) = 0$, this guarantees that $\{W$ an atom of Ω such that W_C entails $X\} = \sum_W P(W)P(W^*_C/W_C) = \sum_W P(C \;\square\!\!\rightarrow X)$.

As we have seen, Stalnaker's Assumption is untenable. Fortunately, Peter Gardenfors has been able to define a more general form of imaging that makes sense even when the assumption fails.[27] When there is no unique world that is "closest" to W in which C is false Gardenfors allows W's probability to be shifted to more than one world, with greater amounts being shifted onto worlds that are more like W. The case he specifically considers assumes a "circle" of worlds that are equally like W and more like W than any world outside the circle. It is fairly easy, however, to see how to extend this scheme to the general case. Start by assuming that for each condition C and world W there is a probability $\rho^C(\bullet, W)$ defined on propositions in Ω, with $P^C(C, W) = 1$, which has the following two features:

---

[25] Lewis defined imaging in terms of "opinionated" probability functions rather than conditional probabilities like $P(X/W_C)$. However, since $P(X/W_C)$ will be equal to 1 if X is true in W_C and to 0 otherwise, it just is one of Lewis's opinionated probability functions. It should also be noted that I am being a bit sloppy here and assuming that the atoms of Ω all have positive probability. In cases where this is not so, the written expression would have to be replaced by an integral involving a probability density function.

[26] Lewis (1976, p. 311).

[27] See Gardenfors (1988, pp. 108–18).

197

- $\rho^C(X, W) > \rho^C(Y, W)$ whenever every world that entails $X \& C$ is more like W than any world that entails $Y \& C$.
- $\rho^C(X, W) = 1$ if it is impossible to partition X into nonempty, disjoint propositions X_1 and X_2 such that every world that entails X_1 is more like W than any world that entails X_2.

$\rho^C(X, W)$ specifies the *proportion* of W's probability that gets shifted to X in the move from P to P($\bullet \backslash C$). The first property guarantees that worlds more like W get a larger share of W's probability, while the second ensures that all of the probability will be distributed (uniformly) over the "circle" of closest worlds provided that such a circle exists. Notice that $\rho^C(X, W) = P(X/W_C)$ when Stalnaker's Assumption holds since $\rho^C(X, W)$ is 1 if W_C entails X and 0 otherwise. In the general case we can define imaging by setting

General Imaging. $P^C(X) = P(X \backslash C) = \sum_W P(W)\rho^C(X, W)$

This agrees with the definitions of Lewis and Gardenfors whenever the latter apply, but it is applicable in cases where they do not. The proportionality measure $\rho^C(X, W)$, and so the imaging function itself, reflects the supposer's judgments of similarity among worlds. These judgments will *not* depend on how likely she takes these worlds to be; the only place where her subjective probabilities enter the equation is through the $P(W)$ term.

It is, of course, unreasonable to expect a person's judgments of relative similarity among worlds to be sufficiently precise and definite to fix a unique ρ^C function for every condition C. By now it should be clear how to handle issues of this sort: The agent's vague judgments, if they are consistent, will determine a large class of proportionality measures. She is committed to an identity $P^C(X) = x$, an inequality $P^C(X) > P^D(Y)$, and so on, just in case these statements hold relative to *all* measures in the class. While further complications could undoubtedly be added to this story, it is best to leave them aside in the present context and to focus on the ideal case in which the subject's estimates of relative similarity among worlds do determine a unique proportion $\rho^C(X, W)$ of W's probability that gets shifted to X in the move from P to P($\bullet \backslash C$).

An imaging function will always satisfy a strengthened version of the "regularity" principle SUP$_3$. Relative to any relation of similarity for possible worlds the subjunctive conditional $C \square \rightarrow X$ is the disjunction of worlds $W \in \Omega$ for which there exists a unique world W_C that entails X and is most similar to W among worlds that entail C. It follows directly that $\rho^C(X, W) = P(X/W_C) = 1$ when W entails

$C \, \square\!\!\rightarrow X$, and that $\rho^{C}(X, W) = P(X/W_{C}) = 0$ when W entails $C \, \square\!\!\rightarrow \neg X$. Thus, we can write

$$P(X\backslash C) = \sum_{W} P(W \, \& \, (C \, \square\!\!\rightarrow X)) \cdot 1 + P(W \, \& \, (C \, \square\!\!\rightarrow \neg X)) \cdot 0$$
$$+ P(W \, \& \, \neg(C \, \square\!\!\rightarrow X) \, \& \, \neg(C \, \square\!\!\rightarrow \neg X)) \rho^{C}(X, W)$$
$$= P(C \, \square\!\!\rightarrow X) + \sum_{W} P(W \, \& \, \neg(C \, \square\!\!\rightarrow X) \, \& \, \neg(C \, \square\!\!\rightarrow \neg X))$$
$$\rho^{C}(X, W)$$

which makes it clear that $P(X\backslash C) \geq P(C \, \square\!\!\rightarrow X)$. And, since $\rho^{C}(X, W)$ can be no greater than 1, it also follows that $P(\neg(C \, \square\!\!\rightarrow \neg X)) \geq P(X\backslash C)$. Thus, we have established that imaging obeys

SUB. $\quad P(\neg(C \, \square\!\!\rightarrow \neg X)) \geq P(X \backslash C) \geq P(C \, \square\!\!\rightarrow X)$

The reader may recall these inequalities from the previous chapter, where they were interpreted as saying that a person's degree of confidence in X when she supposes C subjunctively must fall between her degree of confidence in X being determinately true if C were true, which is given by $P(C \, \square\!\!\rightarrow X)$, and her degree of confidence in X not being *determinately* false if C were true, which is $P(\neg(C \, \square\!\!\rightarrow \neg X))$. This is thus the Weak Ramsey Test for supposition in the subjunctive mood, with $P(C \, \square\!\!\rightarrow X)$ and $P(\neg(C \, \square\!\!\rightarrow \neg X))$ serving as the "best possible" upper and lower bounds on $P(\bullet \backslash \bullet)$. The subjunctive conditional is thus the appropriate "conditioner" for subjunctive supposition.

Unlike the case of matter-of-fact supposition there is no simple condition like MOF that picks out $P(X\backslash C)$ uniquely. Different similarity relations give rise to different imaging functions. When one has a definite similarity relation in mind it is possible to say more about the nature of $P(\bullet \backslash \bullet)$. As we saw in the last chapter, for example, one natural requirement on the sort of subjunctive supposition that is relevant to the evaluation of causal beliefs would have it that

$$1 - \sum_{x} xP(C \, \square\!\!\rightarrow CH(\neg X) = x) \geq P(X\backslash C)$$
$$\geq \sum_{x} xP(C \, \square\!\!\rightarrow CH(X) = x)$$

so that $P(X\backslash C)$ falls between the agent's estimate of the lowest chance that X might have if C were true and her estimate of the highest chance that X might have if C were true. Other constraints would be appropriate for other similarity relations.

199

The inability to specify imaging completely by some simple condition makes it more difficult to formulate simple qualitative constraints on *subjunctive* conditional likelihood rankings. The best one can do without getting into the details of a similarity relation is to say that any such ranking $(\backslash\backslash.>.\backslash\backslash, \backslash\backslash.\geq.\backslash\backslash)$ should satisfy SUP_1^*–SUP_3^* as well as

SUB (Comparative Version). If $(C \,\square\!\!\rightarrow X\,)\backslash\backslash\mathbf{T}\,.\geq.\,\neg(D \rightarrow \neg Y)\backslash\backslash\mathbf{T}$,

then $X\backslash\backslash C\,.\geq.\,Y\backslash\backslash D$

If $(\backslash\backslash.>.\backslash\backslash, \backslash\backslash.\geq.\backslash\backslash)$ is probabilistically representable, then this condition is necessary (but far from sufficient) for it to be represented by a supposition function $P(\bullet\|\bullet)$ in which each $P(\bullet\|C)$ arises from imaging on C relative to some similarity relation among possible worlds (that does not vary with changes in C).

6.4 RÉYNI–POPPER MEASURES AND THE PROBLEM OF OLD EVIDENCE

The most important difference between matter-of-fact supposition and subjunctive supposition has to do with their differing treatments of propositions that the supposer regards as certainly true. A good part of what it means to suppose in a matter-of-fact mode is that one supposes without changing what one sees as "the facts." This is borne out by the fact that matter-of-fact supposition satisfies

Conservation of Certainties. If $P(X) = 1$ and $P(C) > 0$, then

$P(X\|C) = 1$.

Subjunctive supposition does not obey this principle. Indeed, Peter Gardenfors has given an elegant argument that shows that for any similarity relation among worlds there will be at least one probability P whose image does not conserve certainties.[28]

Given this it may seem reasonable to think that any form of supposition that involves conditioning on zero-probability propositions must be carried out in the subjunctive mode. As I suggested earlier, this is wrong. While lowering the probabilities of certainties is a sure indication of a subjunctive supposition *when the supposed proposition C has nonzero probability*, it is possible to make matter-of-fact suppositions when $P(C) = 0$. These are appropriately represented by the so-

---

[28] Gardenfors (1988, pp. 114–17).

called *Réyni–Popper measures*.[29] A *Réyni–Popper measure for* P relative to a set of conditions C is a map $P(\bullet//\bullet)$ from $\Omega \times C$ into the unit interval that satisfies the following requirements for all $X \in \Omega$ and $C, D \in C$.

RP₁ (*Coherence*): $P(\bullet//C)$ is a countably additive probability on Ω.

RP₂ (*Anchoring*): $P(\bullet//C \vee \neg C) = P(\bullet)$.

RP₃ (*Certainty*): $P(C//C) = 1$.

RP₄ (*Multiplication Axiom*): $P(X//C \& D)P(C//D) = P(X \& C//D)$ as long as $(C \& D) \in C$.

The standard conditional probability $P(X/C) = P(X \& C)/P(C)$ is a Réyni–Popper measure relative to $C = \{C \in \Omega: P(C) > 0\}$, but it is not alone. The Multiplication Axiom differs from the usual rule for computing conditional probabilities in allowing for conditioning on propositions whose probability is 0. For any probability P and any set of conditions C that extends $\{C \in \Omega: P(C) > 0\}$ there is always a Réyni–Popper measure $P(\bullet//\bullet)$ that agrees with $P(\bullet/C)$ when $P(C) > 0$ and yet allows conditioning on the elements of C whose unconditional probability is 0.[30] The use of two slashes "//" always indicates an extended conditional probability of this sort, while a single slash "/" will only be used when $P(C) > 0$.

Réyni–Popper measures are fundamentally different from the imaging probabilities obtained in subjunctive supposition. When $P(C) = 0$ it is incorrect to read $P(X//C)$ as the probability that X *would* have if C *were* true. It is, instead, the probability that X *has* if C *is* true. It is sometimes hard to get one's mind around the idea that a person can think about what the world *is* like if a proposition that she is certain is false is in fact true. What makes this so hard is that in the ordinary case such thinking works by straight conjunction. When we suppose a condition C that we take to have some chance of being true in the matter-of-fact mode we just conjoin its content to the content of our current beliefs and renormalize so as to move from P to $P(\bullet/C)$. This

---

[29] The literature on Réyni–Popper measures is vast. It begins with Réyni (1955) and Popper (1959, app. *ii–*v). Important recent discussions may be found in Harper (1976) , Spohn (1986), McGee (1994), van Fraassen (1996), and Hammond (1996).

[30] Popper's official treatment assumed that C would contain every proposition assigned positive probability conditional on any member of C, so that $C \in C$ and $P(X//C) > 0$ implies that $X \in C$. Following van Fraassen (1995, p. 355) we might say that X is an *a priori truth with respect to* P just in case $P(X//C) = 1$ for all $C \in C$. Popper thus required the complement of C to be the set of *a priori* falsehoods with respect to P. I make no such requirement.

cannot be how things go here. If we imagine a Réyni–Popper measure satisfying an extended version of MOF that applies even when $P(C) = 0$ then, since $P(C \& X) = P(C \& Y)$ for all X and Y when $P(C) = 0$, we would get the absurd result that $P(C/C) = P(\neg C/C)$. Thus, the sort of matter-of-fact supposition that is captured by a Réyni–Popper measure is not determined in any simple way by one's antecedent beliefs involving conjunctions. Rather, it is revealed by the fact that one's subjective probabilities obey the following conditionalized version of MOF:

$\textbf{MOF}^{+}$.   If $P(C//D) > 0$ (and even if $P(C) = P(D) = 0$) then

$P(X//C \& D) \geq P(Y//C \& D)$ if and only if $P(C \& X//D)$

$\geq P(C \& Y//D)$.

The appropriate qualitative condition for ensuring that a conditional likelihood ranking will be represented by a Réyni–Popper measure is

$\textbf{MOF}^{+}$ (Comparative Version): If $C//D .>. \neg \textbf{T}//D$, then $X//(C \& D)$

$.>. Y//(C \& D)$ if and only if $(C \& X)//D .>. (C \& Y)//D$

Réyni–Popper measures also satisfy all of the following laws, each of which has an exact analogue for ordinary conditional probabilities:

6.3a (*Weak Ramsey Test*). $P(C \supset X//D) \geq P(X//C \& D) \geq P(X \& C//D)$ as long as $(C \& D) \in \textbf{\textit{C}}$. Moreover, the material conditional and conjunction are, respectively, the logically weakest and logically strongest operations with these properties.

6.3b (*Bayes's Law*). $P(X//C \& D)P(C//D) = P(C//X \& D)P(X//D)$ whenever $(C \& D)$, $(X \& C)$ are in $\textbf{\textit{C}}$.

Proof Hint: Given RP_1–RP_3, the Multiplication Axiom is equivalent to Bayes's Law. It obviously entails Bayes's Law. For the converse, replace X in Bayes's Law by $(X \& C)$.

6.3c (*The Law of "Total Probability"*).

$P(X//D) = P(C//D)P(X//C \& D) + P(\neg C//D)P(X//\neg C \& D)$

6.3d (*Conditional Conservation of Certainty*). If $P(X//D) = 1$ and $P(C//D) > 0$, then $P(X//C \& D) = 1$.

These are all properties that imaging functions typically lack. Indeed, if we apply the Gardenfors result mentioned a few paragraphs back to each $P(\bullet//D)$ we have a rigorous proof of the claim that the sort of supposition represented by Réyni–Popper measures is not in any sense subjunctive. It is, rather, the appropriate extension of matter-of-fact supposition to the case of 0 probability conditions. I want to emphasize this point strongly because it is often assumed that any form of probabilistic belief revision that involves "raising the dead" by increasing the probabilities of certainly false propositions must involve counterfactual beliefs. This is not so. It is logically consistent both to be certain that some proposition is false and yet to speculate about what the world is like if one is *in fact* wrong. To be subjectively certain of something is, after all, not the same as regarding oneself as infallible on the matter. The key point is that wondering how things *are* if one is wrong is not the same as wondering how things *would be* if one *were* wrong.

Even if it is consistent, I suspect that many readers will be wondering why it is even worth bothering about matter-of-fact supposition for propositions that the supposer regards as having no chance of *being* facts. Even granting that it makes sense to speak about conditional probabilities defined for conditions that are certain not to be realized, one might still ask why we should take the time to do so. And, perhaps more importantly, one might question whether such talk really makes sense. Where are these strange conditional probabilities supposed to come from? They aren't ratios of the form $P(X \& C)/P(C)$, of course, since these are undefined. So what are they? It may appear, then, that by bringing Réyni–Popper measures into the discussion I have merely introduced an obscure formal device that has no clear role to play in our epistemic lives.

Not so. We must appeal to matter-of-fact suppositions involving certainly false conditions, the kind of suppositions Réyni–Popper measures capture, to make sense of certain kinds of judgments of *evidential relevance*. The best way to see why this is so is by considering the *problem of old evidence*.[31] On standard Bayesian analyses of evidential relevance, C counts as evidence for X just in case $P(X/C) > P(X)$, and the difference $\mathbf{R}_p(X, C) = P(X/C) - P(X)$ is often used as a measure of *the degree to which C counts as evidence for or against X relative to* P.[32] As Clark Glymour has noted,[33] however, this makes it

---

[31] The problem was introduced by Clark Glymour in his (1980), and it has been discussed in Garber (1983), Jeffrey (1983), Eells (1985), as well as a host of other places. An excellent survey of recent work on the problem of old evidence may be found in Earman (1992).

[32] The following remarks apply equally well to the "relevance quotient" $P(X/C)/P(X)$ and the difference $P(X \& C) - P(C)P(X)$, which have also been proposed as measures of evidential relevance. All three of these functions go to 0 as $P(C)$ goes to 1.

[33] Glymour (1980, Chapter 3).

impossible for any proposition of subjective probability 1 or 0 to be evidence for anything since $P(X/C)$ is $P(X)$ when $P(C) = 1$ and is undefined when $P(C) = 0$. This is clearly the wrong result. A person can be certain that a given statement is true and yet still use it as evidence for some further proposition, as when Einstein took the anomalous advance of the perihelion of Mercury to be evidence for his theory of relativity. It may seem a solution to take Abner Shimony's advice and never adopt an attitude of certainty toward any contingent proposition.[34] But this does not help since, as John Earman notes, the standard measure of evidential relevance allows highly probable propositions very little confirming power, and this too seems wrong.[35] Thus, the simple equation of "E is evidence for H" with "conditioning on E raises H's probability" is inadequate when one is dealing with evidence of extremely high (or low) probability.

There are three separate issues that need to be addressed in connection with this problem:[36]

The Problem of New Hypothesis. How should a rational agent assign an *initial* subjective probability to a newly formulated hypothesis in light of the evidence she already has?

The Problem of Logical Learning.[37] How should a less than ideally rational agent revise her *existing* subjective probability for a hypothesis when she discovers the logical fact that it *entails* some piece of evidence she already has?

The Problem of Evidential Relevance. How can a rational agent, who has a well-defined subjective probability for some hypothesis and who knows what it entails about her evidence, make sense of evidential relationships between the hypothesis and evidence to which she assigns a subjective probability of 1?

These first two problems are hard, and I do not pretend to be able to solve them. The third, however, can be resolved once we realize that a rational believer may have a well-defined conditional probability for X given C even when she is certain that C is false.

To see how this is possible, start by noting that there is an ambiguity in the Bayesian notion of evidential relevance even when $P(C) > 0$.

---

[34] See Shimony (1994).

[35] See Earman (1992, p. 121).

[36] This way of dividing up the issues owes something to Eells (1985) and to Earman (1992).

[37] The best work to date on this problem is found in Garber (1983). Garber's proposal is endorsed and augmented in Jeffrey (1983). It is criticized in Earman (1992), which also serves as the best existing survey of the topic.

All Bayesians agree that the degree to which C counts as evidence for or against X for a given person is a matter of the extent to which learning C would increase or decrease her confidence in X. The ambiguity here has to do with the choice of a reference point from which such changes are measured. The standard view treats them as differences between the agent's current probability $P(X)$ and the probability that she should have were she to learn C, which all Bayesians agree is $P(X/C)$. Among other things, this makes the degree of "confirming power" of a piece of evidence a function of its probability. For any X, $P(\neg C) \geq \boldsymbol{R}_\mathrm{P}(X, C) \geq -P(\neg C)$, which means that $\boldsymbol{R}_\mathrm{P}(X, C)$ goes to 0 as $P(C)$ goes to 1. Thus, the "confirming power" of a highly probable proposition will always be very low, and that of a certainty will be null.

There is, however, another way of looking at the concept of "confirming power." It involves comparing the beliefs that an agent should have upon learning C to those she should have upon learning its negation. This sort of comparison is expressed quantitatively by the difference $\boldsymbol{Q}_\mathrm{P}(X, C) = P(X/C) - P(X/\neg C)$.[38] The function $\boldsymbol{Q}_\mathrm{P}$ provides us with a scale for measuring evidential relevance that does not depend on the unconditional probability of C. $\boldsymbol{Q}_\mathrm{P}(X, C)$ need not go to 0 as $P(C)$ goes to 1, and it is not generally true that more likely propositions have less "confirming power" than less likely propositions. It is easy to show that $\boldsymbol{R}_\mathrm{P}(X, C) = P(\neg C)\boldsymbol{Q}_\mathrm{P}(X, C)$, which means that $\boldsymbol{Q}_\mathrm{P}(X, C)$ is just $\boldsymbol{R}_\mathrm{P}(X, C)$ with the effects of C's probability factored out.

The difference between these two ways of thinking about evidential relevance can be brought out by an example. Imagine that the semifinal round of a chess tournament is about to begin. Abby is pitted against Bill, and Claire is against Dan. Each contestant has played each of the others a hundred times. Here are the results:

	Abby loses	Bill loses	Claire loses	Dan loses
Abby wins.	–	80/100	95/100	95/100
Bill wins.	20/100	–	10/100	90/100
Claire wins.	5/100	90/100	–	10/100
Dan wins.	5/100	10/100	90/100	–

Abby is clearly the class of the field. Her probability of winning it all is 0.76. Claire is obviously the dark horse since her probability of winning is only 0.022. Now, let W be the proposition that a woman wins the tournament ($P(W) = 0.782$), and ask yourself whether learn-

---

[38] Or the quotient $P(X/C)/P(X/\neg C)$, or $P(C)P(\neg C)[P(X \& C) - P(X \& \neg C)]$.

205

ing A = "Abby beats Bill" or C = "Claire beats Dan" would provide you with better evidence for W. From one point of view learning A will not tell you much since you are already confident that Abby will beat Bill and have factored this into your beliefs. Learning C, on the other hand, seems better evidence for W since it would mean a final of either Abby versus Claire, in which case a woman is sure to win, or of Bill versus Claire, in which case a woman has a ninety percent chance of winning. These intuitions reflect the kind of evidential relevance that is measured by $\boldsymbol{R}_\text{P}(X, C) = \text{P}(X/C) - \text{P}(X)$. The numbers work out like this: $\boldsymbol{R}_\text{P}(W, A) = 0.168 < \boldsymbol{R}_\text{P}(W, C) = 0.198$. On the other hand, A is clearly better evidence for W than C is if we have in mind the difference between learning a proposition and learning its negation. If Abby beats Bill then a woman is almost sure to win no matter what happens in the other match. If she loses to Bill, the only way for a woman to win would be for Claire to first beat Dan and then beat Bill, a highly unlikely scenario. In contrast, the difference between learning that Claire beats Dan and learning that she loses to him is not all that striking since Abby is likely to win the final either way. The $\boldsymbol{Q}_\text{P}(X, C) = \text{P}(X/C) - \text{P}(X/\neg C)$ measure bears these intuitions out since $\boldsymbol{Q}_\text{P}(W, A) = 0.86 > \boldsymbol{Q}_\text{P}(W, C) = 0.23$. Here, then, we have a case in which looking at the evidence in the ways described by $\boldsymbol{R}_\text{P}$ and $\boldsymbol{Q}_\text{P}$ leads one to different conclusions about the comparative evidential import of propositions.[39]

Many philosophers have defended $\boldsymbol{R}_\text{P}$ as the correct measure of evidential relevance. $\boldsymbol{Q}_\text{P}$ has had fewer champions.[40] My view is that there is no choice to be made here; it is not as if one of these measures is right and the other wrong. The $\boldsymbol{R}_\text{P}$ and $\boldsymbol{Q}_\text{P}$ measures capture contrasting, but equally legitimate, ways of thinking about evidential relevance, which serve somewhat different purposes. The main contrast between them has to do with their behavior under changes in belief induced by learning. Consider a case in which $\text{P}(C)$ is neither 1 nor 0 and suppose the agent has a *learning experience* whose only immediate effect is to alter her degree of belief in C. Bayesians model such experiences as shifts from the "prior" P to a posterior probability P* that assigns the same probabilities conditional on C and $\neg C$, so that

[39] $\boldsymbol{R}_\text{P}$ and $\boldsymbol{Q}_\text{P}$ will always agree about the qualitative claim that C is evidence in favor of X, that is, $\boldsymbol{R}_\text{P}(X, C) > 0$ iff $\boldsymbol{Q}_\text{P}(X, C) > 0$. What they disagree about is the *degree* to which C is evidence for X, and about comparative claims of the form "C is better evidence for X than C^* is for X^*."

[40] Indeed it was only as this book was going to press that I learned of an excellent forthcoming paper by David Christensen, Christensen (19••), in which $\boldsymbol{Q}_\text{P}$ is defended as the correct measure of evidential relevance.

$P(X/C) = P^*(X/C)$ and $P(X/\neg C) = P^*(X/\neg C)$ for all X. (We will see that this conception of learning as belief revision that leaves conditional probabilities invariant has a reasonable interpretation in terms of "minimality" principles.) In easy to show that every such learning process results in a posterior P^* that is a *Jeffrey shift* of the prior P, so that

$$\textbf{(J)} \quad P^*(X) = p^*P(X/C) + (1 - p^*)P(X/\neg C)$$

where $p^* \in [0, 1]$ is the new probability for C. The following little theorem describes the relevant difference in the behavior of $\textbf{\textit{R}}_P$ and $\textbf{\textit{Q}}_P$ under learning (the proof, which is a straightforward consequence of the definitions, is left to the reader):

Theorem 6.4. *If $P(C) > 0$, $P^*(X) = p^*P(X/C) + (1 - p^*)P(X/\neg C)$ for some $0 < p^* < 1$, then $\textbf{\textit{R}}_{P^*}(X, C) = [P^*(\neg C)/P(C)]\textbf{\textit{R}}_P(X, C)$ and $\textbf{\textit{Q}}_{P^*}(X, C) = \textbf{\textit{Q}}_P(X, C)$.*

This tells us that C becomes less (more) relevant to X in the $\textbf{\textit{R}}_P$ sense whenever the agent undergoes a learning experience whose immediate effect is to raise (lower) her subjective probability for C. Good evidence, in other words, turns into poor evidence as its probability grows, and fair evidence turns into good evidence as its probability shrinks. This contrasts sharply with $\textbf{\textit{Q}}_P$, which is invariant under learning. $\textbf{\textit{Q}}_P$ thus characterizes a sense of evidence for which changing one's degrees of belief in C and $\neg C$ does not lead one to reevaluate one's views about C's confirming power.

What we have here is two related, but still distinct, notions of evidential relevance that behave differently under learning. Not properly distinguishing them is what causes people to think that the Problem of Evidential Relevance poses a serious threat to Bayesian analyses of confirmation. On its face, Theorem 6.4 may not seem pertinent to this matter since it is subject to the qualification $0 < p^* < 1$. I want to suggest, however, that these strict inequalities ought to be weakened. It is possible to view a learning experience that moves an agent's subjective probability for C from $P(C) < 1$ to $P^*(C) = 1$ as a continuous process in which her opinions go through a series of Jeffrey shifts $P_t(C) = p_t P(X/C) + (1 - p_t)P(X/\neg C)$ over a (brief) time interval $[0, 1]$ where $p_t \to 1$ as $t \to 1$, and $P_0(\bullet) = P(\bullet)$ and $P_1(\bullet) = P^*(\bullet)$. Theorem 6.4 then entails that

$$\lim\nolimits_{t \to 1} \textbf{\textit{Q}}_{Pt}(X, C) = P(X/C) - P(X/\neg C)$$
$$\lim\nolimits_{t \to 1} \textbf{\textit{R}}_{Pt}(X, C) = 0$$

These identities highlight a crucial difference between $\boldsymbol{R}_\mathrm{P}$ and $\boldsymbol{Q}_\mathrm{P}$. While the lower limit is of no interest since it is completely insensitive to differences between X and C, the upper one gives us a way of cashing out relations of evidential relevance involving propositions with subjective probabilities of 0 or 1. Since $\boldsymbol{Q}_{\mathrm{P}t}(X, C)$ remains fixed no matter how close t gets to 1, it would be gratuitous not to suppose that it stays fixed at $t = 1$. The only ground for denying this would be some version of the "certainty fallacy," which portrays the move that takes C's probability from $1 - \varepsilon$ to 1 as fundamentally different in kind from a move that takes it from $1 - 2\varepsilon$ to $1 - \varepsilon$ no matter how small ε may be.[41] It is hard to see, however, why certainty should make any difference in this context. Does it seem reasonable that a person who feels that C is better evidence than $\neg C$ for X at $P(C) = 0.9, 0.99,$ $0.999, \dots$ should suddenly cease making this judgment when $P(C)$ reaches 1? A more plausible picture will treat her views as remaining stable under learning experiences involving C, *even those that push C's probability to 1 or 0*. The strict inequalities in Theorem 6.4 should be weakened to take this stability into account.

The same reasoning extends straightforwardly to the conditional probabilities of C and $\neg C$. Since these remain fixed under nonextreme learning experiences involving C we should regard them as remaining intact under extreme learning experiences as well, so that $P(X/C) = P^*(X/C) = P^*(X)$ and $P(X/\neg C) = P^*(X/\neg C)$ when $P(C) = 1$. Unfortunately, this third identity is nonsense. The only way to get a sensible value for it is to suppose that a person's beliefs at the instant C becomes certain are given by a Réyni–Popper measure in which $P^*(X//\neg C) = \lim_{t \to 1} P_t(X/\neg C)$. Hence, if we want to capture those features of doxastic states that remain invariant under learning, as we definitely do, then we need to think of such states as being characterized by Réyni–Popper measures and to understand learning as a process whereby experience transforms a "prior" $P(\bullet//\bullet)$ into a "pos-

---

[41] I can imagine one reason for thinking that the change from uncertainty to certainty is different in kind from any change involving uncertainties. Since certainty is an *irrevocable* belief state (once $P(C) = 1$, further learning can never change this fact), one might argue that the amount of evidence it should take to raise a belief from $1 - \varepsilon$ to 1 should always be greater than that required to raise it from $1 - 2\varepsilon$ to $1 - \varepsilon$. While there might be something to this objection, it does not affect the point at hand. The only consistent way to defend the claim that it is always harder to raise a belief from $1 - \varepsilon$ to 1 than from $1 - 2\varepsilon$ to $1 - \varepsilon$ is by treating subjective certainty (for contingent propositions) the way the special theory of relativity treats the speed of ligh – as an unattainable limit – so that no amount of evidence should ever be sufficient to raise the probability of a contingent proposition to 1. If this is the case, then the zero-probability problem is not going to arise.

terior" $P*(\bullet//\bullet)$, where the two measures are defined over the same set of conditions $\mathbf{C}$.

If it makes sense to model learning in this way, then a solution to the Problem of Evidential Relevance is at hand. A person's state of opinion at any time will be a Réyni–Popper measure $P(\bullet//\bullet)$ whose associated set of conditions $\mathbf{C}$ contains all propositions to which she has *ever* assigned positive probability. For any $C \in \mathbf{C}$, the quantity $\mathbf{Q}_P(X, C) = P(X//C) - P(X//\neg C)$ will then be well defined, and its value will not depend on what the probabilities for C and its negation happen to be. This allows us to capture the sense in which C can be evidence for X even for a person who is certain C is true (or false). While C cannot be evidence in the $\mathbf{R}_P$ sense when $P(C) = 1$ (or 0), it can still be evidence in the $\mathbf{Q}_P$ sense; an agent can be quite certain of C (or $\neg C$) and yet think that C is evidence for X in the sense that X is more likely given C than given $\neg C$. Judgments of this sort are what underlie all our intuitions about the confirming power of "old" evidence. The fact that Réyni–Popper measures allow us to understand evidential relations of this sort is, in my view, a conclusive reason for using them to model beliefs.[42]

The one fly in the ointment here is an argument, recently put forward by John Earman, that purports to show that a version of the Problem of Evidential Relevance arises even for the $\mathbf{Q}_P$ measure of evidential relevance.[43] Philosophers trying to understand how hypotheses are tested in science are most interested in cases in which some hypothesis X is confirmed by a piece of evidence C that it logically entails. Since $P(X/\neg C) = 0$ in this situation it will always be true that $\mathbf{Q}_P(X, C) = P(X/C)$ and thus that $\mathbf{Q}_P(X, C) = P(X)$ when $P(C) = 1$. According to $\mathbf{Q}$, then, all statements of probability 1 that X entails will confirm X to degree $P(X)$. It follows that if C and D are both entailed by X, then learning experiences that raise the subjective probabilities of both to 1 will "wash out" any differences between the two conditions as far as their ability to confirm X is concerned.[44] This, according to Earman, shows that $\mathbf{Q}$ "is not a suitable measure of evidential relevance."[45]

---

[42] The solution to the Problem of Evidential Relevance I am offering here is a version of the general strategy of "giving a probability a memory." See Skyrms (1983).

[43] Earman (1992, pp. 120–21).

[44] As David Christensen has observed, Earman's misgivings about $\mathbf{Q}_P$ have nothing special to do with the fact that C and D are believed with certainty. The problem arises whenever $P(C)$ and $P(D)$ are close to 1 because $\mathbf{Q}_P(X, C) - \mathbf{Q}_P(X, D)$ goes to 0 as $P(C \,\&\, D)$ goes to 1. See Christensen (unpublished manuscript).

[45] Earman (1992, p. 121).

Earman does not say why he takes this sort of washing out to be problematic, but it is easy enough to imagine what might be bothering him. Consider the following statements:

X: All ravens are black.
C: All ravens observed before today were found to be black.
D: The first raven observed today will be found to be black.

Clearly, *C*'s truth confirms *X* much more strongly than *D*'s truth does. Moreover, this difference in confirming power does not seem to depend on how strongly any of the three propositions are believed; even a person who is certain of both *C* and *D* should be able to recognize that *C* provides much, much better evidence for *X* than *D* does. Since $\mathbf{Q}$ represents *C* and *D* as confirming *X* equally when P(*C*) and P(*D*) are equal to 1 it seems unable to capture this obvious fact about confirmation. Even worse, when P(*C*) = P(*D*) = 1 their confirming power relative to *X* coincides with that of *C* & *D*. To see why this is problematic write out *C* as (*C* & *D*) ∨ (*C* & ¬*D*) and notice that it is composed of a disjunct, *C* & *D*, whose truth would confirm *X* and a disjunct, *C* & ¬*D*, whose truth would conclusively falsify *X*. Given this, it seems clear that *C* should confirm *X* *less* strongly than *C* & *D* does. Since $\mathbf{Q}$ assigns them equal confirming power it cannot be a suitable measure of evidential relevance, or so Earman might argue.

We should not be convinced. The right way to think about the issue, as David Christensen has observed,[46] is to consider the way in which learning-induced changes in *C*'s probability can change the value of $\mathbf{Q}_p(X, D)$ by altering P(*X/D*). While the value of $\mathbf{Q}_p(X, D)$ is invariant under changes in *D*'s probability it *can* vary with changes in the probabilities of *other propositions*. A Jeffrey shift involving *C* of form P*(•) = *p**P(•/*C*) + (1 − *p**)P(•/¬*C*) will alter $\mathbf{Q}_p(X, D)$ so that

$$\mathbf{Q}_{p*}(X, D) = P*(X/D) = P*(X)/P*(D)$$
$$= p*P(X/C)/[p*P(D/C) + (1 - p*)P(D/\neg C)].$$

As *p** increases the rightmost expression converges to P(*X/D* & *C*), which means that, according to $\mathbf{Q}$, *D*'s evidential relevance *vis-à-vis X* increasingly approximates that of *D* & *C* as *C*'s probability increases. Relative to $\mathbf{Q}$, then, the extent to which *D* counts as evidence in favor of *X* for a given person depends on that person's degree of belief for *C*. More generally, the extent to which one proposition confirms or

---

[46] Christensen (unpublished manuscript).

disconfirms another for a person is a function of the strengths of that person's beliefs about other propositions.

This is exactly how it should be. The one fact about evidential relevance that has been established beyond doubt in recent years is that confirmation is *holistic*. The extent to which one proposition provides evidence for another is not an absolute, belief independent matter; it varies from person to person depending on their background state of opinion. This holism is reflected in Bayesian accounts of confirmation through the dependence of an agent's probabilities conditional on D on changes in her unconditional probabilities of propositions other than D. Both $\boldsymbol{Q}$ and $\boldsymbol{R}$ incorporate this dependence by forcing D's confirming power to vary with changes in the probabilities of other propositions. The main difference between the two measure is that $\boldsymbol{R}$ allows D's confirming power to vary with changes in its own probability, whereas $\boldsymbol{Q}$ does not.

Once one recognizes that confirmation has this holistic character it becomes clear that Earman's misgivings about $\boldsymbol{Q}$ are groundless. The extent to which D is evidence for X for a person *should* be a function of her degree of belief in C. As C's probability approaches 1, thus forcing the probability of D & $\neg C$ closer and closer to 0, D *does* become increasingly better evidence for X in the sense of evidence that $\boldsymbol{Q}$ is intended to capture. Since X's unconditional probability and its probability conditional on D always increase as $P(C)$ increases, every upward movement in the latter endows D with the ability to falsify a slightly more probable proposition and to raise that proposition's unconditional probability to a slightly higher level. Thus, the difference between learning D and learning $\neg D$ becomes increasingly significant as C becomes more and more certain. This is precisely the sense of evidential relevance that $\boldsymbol{Q}$ was meant to characterize. When evidence is thought of in this way it is not at surprising that all propositions that entail X and have probability 1 end up confirming X to the same degree when they are believed with certainty. After all, every proposition that entails X has the same (maximal) power to falsify X and its ability to raise X's probability is exhausted once it is believed to the maximum degree. Given this, the result that Earman finds troubling should not trouble us at all. Given the sense of "is evidence for" that $\boldsymbol{Q}$ is supposed to capture, it is exactly what we should expect and want.

But, one might wonder, does $\boldsymbol{Q}$ capture the sense of evidential relevance that is most important to our understanding of confirmation? It captures one of them, $\boldsymbol{R}$ captures another, but it must be admitted that there are features of the concept of evidence that neither $\boldsymbol{Q}$ nor $\boldsymbol{R}$ adequately explains. If one thinks about the raven case it should be clear that even a person who believes C and D to degree 1 can recognize a sense in which C is better evidence than D is for X; he can recognize that his subjective probability for X would be decreased far more by an "anti-learning" experience that somehow lowered $P(C)$ to some value intermediate between 1 and 0 than by an experience that lowered $P(D)$ by the same amount. There is, I think, a sense of the term "evidence" in which C is counted as providing better evidence than D does for X just when a change in C's probability would produce a *larger* effect on X's probability than an identical change in D's probability would produce. Let's call this the *incremental* sense of evidential relevance. When $P(C)$ assumes an intermediate value we can capture (two aspects) of this incremental sense of relevance using "normalized" versions of $\boldsymbol{Q}$ and $\boldsymbol{R}$ defined, respectively, by $\boldsymbol{Q}'_{\mathrm{P}}(X, C) = \boldsymbol{Q}_{\mathrm{P}}(X, C)/(1 - P(C))$ and $\boldsymbol{R}'_{\mathrm{P}}(X, C) = \boldsymbol{R}_{\mathrm{P}}(X, C)/(1 - P(C))$. When $P(C) = 1$ neither of these quantities is defined. As the raven case shows, however, we need a way of making sense of C's incremental power to confirm X even when $P(C) = 1$. Thus, there is still work to be done before we arrive at a fully adequate characterization of the notion of evidential relevance in all its many guises. If this was Earman's underlying point, then I will happily concede it, but with the proviso that $\boldsymbol{Q}$ captures one perfectly good sense of evidential relevance and that the sense it captures is the one that is especially relevant to the Problem of Evidential Relevance being treated here.

Attentive readers will have noticed at least one flaw in this pretty picture. My solution to the Problem of Evidential Relevance and my response to Earman only work if there is some natural way of characterizing learning for belief systems represented by Réyni–Popper measures. While an agent's views about the $\boldsymbol{Q}_{\mathrm{P}}$ evidential relevance of C to X remain fixed when she learns about C, they might well change in other types of learning experiences. If she has an experience whose direct effect is to alter her probability for some other proposition D, for example, her conditional probabilities for X given C and given $\neg C$ might change, and this may shift $\boldsymbol{Q}_{\mathrm{P}}(X, C)$ as well. For the position I have been defending to be at all plausible it must be possible to say something definitive about the "kinematics" of such changes. More generally, we need to be able to explain how learning is to be modeled in the context of Réyni–Popper measures. The question is how $P(X//C)$ will change when the agent has a learning experience involving some proposition D other than C.

We can start by getting clear about why a simple application of

Jeffrey's rule will not suffice to answer this question. Jeffrey's rule tells us that the learner's new "anchor" probability will be $P^*(X) = P^*(X//T) = p^*P(X//C) + (1 - p^*)P(X//\neg D)$ after a learning experience whose immediate effect is to move D's probability from some value between 0 and 1 to $P^*(D) = p^*$. As the following lemma indicates, this does convey some useful information about what the rest of $P^*(\bullet//\bullet)$ looks like.

Lemma 6.5. *Let $P(\bullet//\bullet)$ and $P^*(\bullet//\bullet)$ be Réyni–Popper measures defined on $\Omega \times \mathbf{C}$. Suppose that P^*'s "anchor" probability is a Jeffrey shift of P's, so that $P^*(\bullet) = P(\bullet//T) = p^*P(\bullet//D) + (1 - p^*)P(\bullet//\neg D)$ for some condition $D \in \mathbf{C}$ such that $1 > P(D) > 0$ and some $p^* \in [0,1]$. Then, for any $C \in \mathbf{C}$ such that both $P(C \& D)$ and $P(C \& \neg D)$ are both positive, one will have*

(a) $P^*(X//C \& D) = P(X//C \& D)$ and
      $P^*(X//C \& \neg D) = P(X//C \& \neg D)$

(b) $P^*(X//C) = \dfrac{P^*(D)P(X \& C//D) + P^*(\neg D)P(X \& C//\neg D)}{P^*(D)P(C//D) + P^*(\neg D)P(C//\neg D)}$

   $= P^*(D//C)P(X//C \& D) + P^*(\neg D//C)P(X//C \& \neg D)$

(c) $P^*(X//C) = \dfrac{P^*(D)P(C//D)}{P^*(D)P(C//D) + P^*(\neg D)P(C//\neg D)}$

   $= \dfrac{P^*(D)[P(D//C)/P(D)]}{P^*(D)[P(D//C)/P(D)] + P^*(\neg D)[P(\neg D//C)/P(\neg D)]}$

(Again, the proof may be found in the Appendix to this chapter.) These results are good as far as they go, but since 6.5a–6.5c have only been shown to hold when $P(C \& D)$ and $P(C \& \neg D)$ are both nonzero, they tell us nothing about how learning works in true Réyni–Popper measures; since C's probability is nonzero their "//"s can all be replaced by "/".

The key to extending the Jeffrey model of learning so that it can be applied when $P(C) = 0$ lies in 6.5a and the last identity of 6.5c, both of which I propose as constraints on the learning process within the Réyni–Popper framework. 6.5a is a natural extension of the basic Bayesian insight that an experience that directly changes a person's subjective probabilities for D will leave ratios of probabilities of con-

213

junctions involving D intact. One consequence of this in the case of ordinary conditional probabilities is that $P^*(X/Y) = P(X/Y)$ whenever X and Y entail D. 6.5b simply makes this good when Y is a (consistent) proposition that entails D and has a prior unconditional probability of 0. 6.5c is a slightly more complicated variation on the same basic theme. It shows how the relation in its first identity, which is undefined when $P(C) = 0$, can be preserved when $P(D//C)$ and $P(\neg D//C)$ make sense.

We can use 6.5a and the second equation of 6.5c to secure the following extension of Jeffrey's rule for Réyni–Popper measures (after a bit of algebra):

$$(\mathbf{J+})\; P^*(X//C) = \frac{\lambda[P(D//C)P(X//C \,\&\, D)] + \mu[P(\neg D//C)P(X//C \,\&\, \neg D)]}{\lambda P(D//C) + \mu P(\neg D//C)}$$

where $\lambda = P^*(D)/P(D)$ and $\mu = P^*(\neg D)/P(\neg D)$. Note that this equation makes sense even if $P(C) = 0$ as long as $P(D//C)$ and $P(\neg D//C)$ are defined, and that it reduces to Jeffrey's equation when $P(C \,\&\, D)$ and $P(C \,\&\, \neg D)$ are both nonzero. J+ fully describes the effect of learning about D on belief states represented by Réyni–Popper measures, thereby making it clear how a person who is already sure that some proposition C is true (or false) can update her views about probabilities conditional on C when she learns something about the other proposition D.

It must be emphasized that J+ holds only when both $P(D)$ and $P(\neg D)$ are nonzero. It is thus silent about *nonmonotonic* learning experiences whose immediate effect is to lower the probability of a proposition that was antecedently believed with certainty. It should be clear, however, that Réyni–Popper measures are well suited for handling nonmonotonic changes in belief induced by learning since, as I have been at some pains to argue, they allow us to make sense of evidential relations involving propositions that have probability 0. There has been a fair amount of work done in this area,[47] and I believe that some version of the Réyni–Popper approach is the best way to handle questions about nonmonotonic learning. While a full discussion of these issues would take us too far afield, they do bring to the fore a question that does needs to be asked: What is the precise nature of the relationship between the kind of belief revision that is involved

---

[47] Some important references here are Harper (1975), Levi (1980), Sphon (1986), Gardenfors (1988), Hammond (1994), and van Fraassen (1995).

in supposing in the matter-of-fact mode and the kind of belief revision involved in learning? This is the subject of the next section.

At a certain level it is easy to characterize the relationship between matter-of-fact supposition and learning. Since a person who supposes C in the matter-of-fact mode will revise her beliefs by moving from P to $P(\bullet/C)$ when $P(C) > 0$, and since a learning experience whose immediate effect is to raise C's probability to 1 also forces her to move from P to $P(\bullet/C)$, the two processes are the same. Someone who supposes C in the matter-of-fact mode revises her opinions as she would if she were to learn that C is true. End of story? Not quite. The relationship between matter-of-fact supposition and learning is much more subtle than this. To understand it, we must move to the more general perspective that we took in the first section of this chapter and ask about the *similarity gauge* that underlies supposition in the matter-of-fact mode.

Recall that a similarity gauge $\sigma(P, Q)$ is a measure of the extent to which a pair of probability functions defined on Ω "diverge" from one another, and that each such gauge defines a belief revision rule that, for any prior probability P and convex set of probabilities Φ, selects a final probability $P(\bullet\|\Phi)$ that minimizes $\sigma(P, Q)$ among all the probabilities in Φ. The question, which I have avoided up to now, is: What is the right similarity gauge for matter-of-fact supposition?

There has been a great deal of work done on this issue.[48] Three candidates for σ have received the most attention:

Kullback–Leibler Information. $\sigma(P, Q) =$

$$\sum\nolimits_{W \in \Omega} Q(W) \log(Q(W)/P(W))$$

Hellinger Distance. $\sigma(P, Q) = \sum\nolimits_{W \in \Omega} \left(Q(W)^{1/2} - P(W)^{1/2}\right)^2$

Variational Distance. $\sigma(P, Q) = \sup\{|Q(X) - P(X)|: X \in \Omega\}$

What makes these interesting is that their associated selection rules are *Bayesian* in the sense that they designate the Jeffrey shift as the most similar probability to P subject to constraints that describe learn-

---

[48] See especially Diaconis and Zabell (1983), Skyrms (1987), Seidenfeld (1986), van Fraassen (1981), Jaynes (1978), and Hobson (1971).

ing, that is, they force it to be the case that $P(\bullet\|\Phi) = \Sigma_j q_j P(\bullet/C_j)$ when $\Phi = \{Q \in \pi: Q(C_1) = q_1 \& \ldots \& Q(C_n) = q_n\}$ and $1 > P(C_j) > 0$ for all j.[49] (The first two gauges are convex, and thus uniquely minimized by the Jeffrey shift, while the third is a nonconvex, "Lewisian" rule, which makes a whole class of functions "equally close" to P.) These results lend credence to the idea that belief revision by learning and belief revision by matter-of-fact supposition are essentially the same process.

Nonetheless, the idea is wrong. To appreciate why, we need to see how supposition and learning work in contexts more general than those to which Jeffrey conditioning applies. For two examples, suppose a three-sided die is about to be tossed, and imagine that your subjective probability P is defined over the algebra generated by the propositions S_j = "The side with j spots comes up" for $j = 1, 2, 3$. Here are two experiences you might have that cannot be adequately modeled in the simple learning-as-change-of-unconditional probability framework that Jeffrey conditionalization envisions.

The "Judy Benjamin" problem (van Fraassen):[50] Your prior probability is $P(S_1) = 1/2$, $P(S_2) = P(S_3) = 1/4$. You learn that your new posterior for S_2 conditional on $(S_2 \vee S_3)$ should be 3/4 (perhaps by being told that 3/4 is the objective of chance of two spots coming up on the die given that two or three will come up). Here, the constraint is $\Phi = \{Q \in \pi: Q(S_2/S_2 \vee S_3) = 3/4\}$. What is the right final probability for you to adopt?

The "Brandeis Dice" problem (Jaynes):[51] Your prior is uniform probability: $P(S_1) = P(S_2) = P(S_3) = 1/3$. You learn that the expected number of spots is 1.5 (perhaps by being told that this was the average value in a long series of random tosses of the die). Here $\Phi = \{Q \in \pi: Q(S_1) + 2Q(S_2) + 3Q(S_3) = 1.5\}$. What is the right final probability?

If God were a Bayesian, then all Bayesian selection rules would agree about cases like these. No such luck! Here are the approximate values we get by using the three rules (with the probabilities for S_3 suppressed):

[49] See Diaconis and Zabell (1983, pp. 822–30). The Zabell and Diaconis result generalizes to cover learning experiences that directly affect the probabilities of propositions in a countable partition.
[50] van Fraassen (1981).
[51] Jaynes (1978).

216

	Judy Benjamin		Brandeis Dice	
	$P(S_1)$	$P(S_2)$	$P(S_1)$	$P(S_2)$
Kullback–Leibler	0.5156	0.3633	0.5482	0.3388
Hellinger Distance	0.6880	0.2340	0.4866	0.3850
Variational Distance	0.4980	0.3765	0.6666	0.2500

These are striking differences. They indicate the Kullback–Leibler, Hellinger, and Variational Distance gauges give rise to very different kinds of belief revision in contexts where Jeffrey conditionalization cannot be used. Since at most one gauge can represent matter-of-fact supposition, we seem to be left with the task of figuring out which, if any, is the correct measure to use. We are also left with the possibility that the right similarity gauge for matter-of-fact supposition does not always lead to belief revisions that are consistent with learning. Indeed, this is exactly what turns out to be the case.

Since the remarks that follow will not depend in any essential way on which semi-Bayesian similarity gauge is used in characterizing matter-of-fact supposition I will assume that Kullback–Leibler information does the job. I should say that I do, in the end, think this is the right rule to use because the policy of choosing Q to minimize $\Sigma_{W \in \Omega} Q(W)\log(Q(W)/P(W))$ can be plausibly portrayed as a formal version of the informal rule that "one should not jump to unwarranted conclusions, or add capricious assumptions, when accommodating one's belief states to the deliverances of experience."[52] I will not, however, present arguments for this conclusion here.

Assuming that Kullback–Leibler information is the right similarity gauge for matter-of-fact supposition, the issue becomes one of determining whether the sort of belief revision that proceeds by minimizing this function is the same as the sort of belief revision that is involved in learning. It turns out that it cannot be. Using an example first presented by K. Friedman and A. Shimony, Brian Skyrms has shown that supposition and learning are fundamentally different processes.[53] Imagine, once again, that a three-sided die is about to be tossed and that your subjective probability P is defined over the algebra generated by the propositions S_j = "The side with j spots comes up." for j = 1, 2, 3. Here is the problem:

The loaded die: The die is loaded and will always come up the same way. You do not know which way it is loaded, however, so your prior probability is uniform over the three possibilities: $P(S_1)$ =

[52] van Fraassen (1981, p. 376).
[53] See Friedman and Shimony (1971) and Skyrms (1987).

$P(S_2) = P(S_3) = 1/3$. Let C say that the average value of the number of spots in a long series of independent trials was 2.

There are two ways to view this piece of information. One can either (i) treat C as imposing a constraint $\Phi = \{Q \in \pi: Q(S_1) + 2Q(S_2) + 3Q(S_3) = 2\}$ on one's posterior probability or (ii) treat it as a proposition that has been learned. In the first case, the policy of minimizing Kullback–Leibler information (or any acceptable measure of divergence among probabilities) will advise you to make *no* changes whatsoever in your beliefs because your prior probability is already in the constraint set. In the second case you would condition on C, and your new probabilities should be $P(S_2/C) = 1$ and $P(S_1/C) = P(S_3/C) = 0$. What this difference in values shows, Skyrms rightly notes, is that the epistemic processes involved in treating information about the expected value of a quantity as a constraint and treating it as something that one has learned are different in kind. They happen to coincide when the quantity in question is the probability of a proposition, or the probabilities of propositions in some partition, but they diverge in more general settings.

Skyrms also correctly points out that minimizing divergence with respect to a constraint is the hallmark of supposition, not of learning. He also seems to have thought, however, that minimizing Kullback–Leibler information is what we do when we suppose in the *subjunctive* sense:

Updating subjective belief to assimilate a given piece of information and supposing what the world would be like were that bit of information true, are distinct mental processes for which different rules are appropriate . . . The difference is often marked by the distinction between the indicative and subjunctive mood. The Warrenite will assert: "If Oswald didn't kill Kennedy, then someone else did" but deny: "If Oswald hadn't killed Kennedy then someone else would have. . . . Much of the debate about [minimizing Kullback–Leibler information] appears to proceed on the assumption, tacit or explicit, that that [it] is an inductive rule, i.e. a rule for *updating* subjective probabilities. I want to suggest that this is the wrong way to look at [it]. Properly viewed [the policy of minimizing Kullback–Leibler information] is a rule for *stochastic hypothesizing*; a rule for *supposing*.[54]

I agree completely with the claim that learning and subjunctive supposition are different, and with the claim that learning and the process captured by maximizing Kullback–Leibler information are fundamentally different. What I deny is that the process captured by maximizing

[54] Skyrms (1987, p. 225).

Kullback–Leibler information has anything to do with *subjunctive* supposition. My reason is simple: The probability Q that minimizes $\Sigma_{W \in \Omega} Q(W) \cdot \log(Q(W)/P(W))$ is always *absolutely continuous* with respect to P in the sense that $P(X) = 0$ implies $Q(X) = 0$. This means the selection function $P(\bullet \| \bullet)$ it defines always obeys Preservation of Certainties, which is a hallmark of matter-of-fact rather than subjunctive supposition. A person who provisionally adopts a subjective probability that maximizes Kullback–Leibler information is always supposing in the matter-of-fact mode.

Now, while I do want make clear that there is a difference between learning and matter-of-fact supposition, I also do not want to overemphasize the point. The processes are closely related. Both can be seen as instances of minimizing divergence among probabilities with respect to a constraint. I even think that the similarity gauge is the same in both cases. (My preferred gauge, as I have said, is Kullback–Leibler information, but I am more confident that there is only one gauge for both processes than I am that this is it.) The difference between the matter-of-fact supposition and learning really has to do with the way in which the relevant constraint set is determined.

The Friedman/Shimony example is instructive here. When one has a piece of data that indicates the expected value of some random variable $\mathrm{Exp}(f) = x$ one can either (i) proceed as a supposer would and choose a constraint that contains all probabilities that assign an expected value of x to f or (ii) proceed as a learner would and choose a constraint that contains all probabilities that assign the proposition that $\mathrm{Exp}(f) = x$ a value of 1. The possibilities, in other words, are $\Phi = \{Q \in \pi: \mathrm{Exp}_Q(f) = x\}$ and $\Phi^* = \{Q \in \pi: Q(\mathrm{Exp}(f) = x) = 1\}$. For this second constraint to make sense, $\mathrm{Exp}(f) = x$ must express a definite proposition (measurable set) that appears in the algebra over which the supposer's subjective probabilities are defined. For $\mathrm{Exp}(f) = x$ to express a definite proposition, however, there has to be a probability function, call it R, relative to which the expectation in question is computed. (In the dice example this probability is the objective bias of the die.) The constraint associated with learning is really $\Phi^* = \{Q \in \pi: Q(\mathrm{Exp}_R(f) = x) = 1\}$.

Now this latter constraint will be a subset, usually a proper subset, of the former provided that each probability Q in Φ^* treats R is an "expert"[55] relative to f in the sense that $\mathrm{Exp}_Q(f/\mathrm{Exp}_R(f) = x) = x$; that is, Φ^* will be contained in Φ if Q's expectation for f conditional on knowing R's expectation for f is always equal to R's expectation for f.

---

[55] The best existing discussion of "expert" probabilities is Gaiffman (1988).

In such a case the difference between supposing and learning comes down to the difference between choosing a new probability that simply meets the constraint Φ and choosing one that meets the constraint *because it assigns probability 1 to the proposition that some expert probability satisfies* Φ. These two operations always coincide when the random variable in question is a proposition (since in this case the relevant expert probability is just the proposition's truth-value), but in other cases they come apart.

While there is a great deal more to be said about the relationship between supposition and learning, we shall have to leave these topics where they stand. There are, admittedly, a number of questions that have been left open. Perhaps the most important one, which I have not even tried to address at all, is that of finding appropriate measures of divergence for subjunctive supposition (given a measure of similarity over possible worlds). This is a very difficult problem, and I have nothing useful to say about it at the moment. Fortunately, one does not need to know how to solve it to make significant headway in the foundations of causal decision theory. This is the task to which we now turn.

6.6 APPENDIX: PROOFS OF THEOREM 6.1, THEOREM 6.2, AND LEMMA 6.5

Theorem 6.1. *If σ is a similarity gauge, then its associated supposition function obeys*

SUP_1 *(Coherence): $P(\bullet\|C)$ is a countably additive probability on Ω.*
SUP_2 *(Certainty): $P(C\|C) = 1$.*
SUP_3 *(Regularity): $P(X\|C) \geq P(X \& C)$ as long as $C \in \boldsymbol{C}$.*

Proof: SUP_1 and SUP_2 are obvious. To establish SUP_3, suppose for purposes of *reductio* that $P(X \& C) > P(X\|C)$ when $C \in \boldsymbol{C}$. We will construct a probability function P* such that $P^*(C) = 1$ whose values fall uniformly closer to those of P than those of $P(\bullet\|C)$ relative to the partition $\{(X \& C), (\neg X \& C), \neg C\}$. This shows that $P(\bullet\|C)$ cannot minimize (P, Q) subject to $Q(C) = 1$ without contradicting Weak Uniform Dominance. Set $\lambda = P(X \& C) - P(X\|C)$ and $\mu = P(\neg X\|C) - P(\neg X \& C)$. Clearly $1 \geq \lambda, \mu > 0$. Define P* by

$$P^*(Y \& X) = (1-\mu)P(Y \& X\|C) + \mu P(Y \& X \& C)$$
$$P^*(Y \& \neg X) = (1-\lambda)P(Y \& \neg X\|C) + \lambda P(Y \& \neg X \& C)$$

and

$$P*(Y) = P*(Y \& X) + P*(Y \& \neg X)$$

Notice that $P*(C) = 1$ because $P*(\neg C \& X) = P*(\neg C \& \neg X) = 0$. $P*$ is a probability because

$$
\begin{aligned}
P*(X) + P*(\neg X) &= (1 - \mu)P(X \| C) + \mu P(X \& C) + (1 - \lambda)P(\neg X \| C) \\
&\quad + \lambda P(\neg X \& C) \\
&= (1 - \mu)P(X \| C) + \mu P(X \& C) + (1 - \lambda)P(\neg X \| C) \\
&\quad + \lambda P(\neg X \& C) \\
&= P(X \| C) + P(\neg X \| C) + \mu[P(X \& C) - P(X \| C)] \\
&\quad + \lambda[P(\neg X \& C) - P(\neg X \| C)] \\
&= 1 + \mu\lambda - \lambda\mu = 1
\end{aligned}
$$

and because for any pairwise incompatible set of propositions $\{Y_1, Y_2, Y_3, \ldots\}$

$$
\begin{aligned}
P*(V_j Y_j) &= P*(V_j(Y_j \& X)) + P*(V_j(Y_j \& \neg X)) \\
&= [(1 - \mu)P(V_j(Y_j \& X) \| C) + \mu P(V_j(Y_j \& X \& C))] \\
&\quad + [(1 - \lambda)P(V_j(Y_j \& \neg X)) \| C) + \lambda P(V_j(Y_j \& \neg X \& C))] \\
&= \left[(1 - \mu)\sum_j P(Y_j \& X \| C) + \mu \sum_j P(Y_j \& X \& C)\right] \\
&\quad + \left[(1 - \lambda)\sum_j P(Y_j \& \neg X \| C) + \lambda \sum_j P(Y_j \& \neg X \& C)\right] \\
&= \sum_j [(1 - \mu)P(Y_j \& X \| C) + \mu P(Y_j \& X \& C)] \\
&\quad + \sum_j [(1 - \lambda)P(Y_j \& \neg X \| C) + \lambda P(Y_j \& \neg X \& C)] \\
&= \sum_j [P*(Y_j \& X) + P*(Y_j \& \neg X)] = \sum_j P*(Y_j)
\end{aligned}
$$

Since $P*(Y \& X) = P*(Y \& X \& C)$ is a mixture of $P(Y \& X \| C)$ and $P(Y \& X \& C)$ it must fall between these two values. Likewise, $P*(Y \& \neg X) = P*(Y \& \neg X \& C)$ must fall between $P(Y \& \neg X \| C)$ and $P(Y \& \neg X \& C)$. And, since $P*(Y \& \neg C) = P(Y \& \neg C \| C) = 0$, this means that $P*(\bullet)$ uniformly dominates $P(\bullet \| C)$ on the partition $\{X \& C, \neg X \& C, \neg C\}$, and thus that $\sigma(P, P(\bullet \| C)) \geq \sigma(P, P*)$. Since $P(\bullet)$ is not identical to $P(\bullet \| C)$, this contradicts the hypothesis that $P(\bullet \| C)$ uniquely minimizes σ subject to the constraint. This completes the proof of Theorem 6.1. ∎

Theorem 6.2. *Except in the extreme case where C entails X or ¬X, if $ and @ are any connectives such that the inequalities P(C $ X) ≥ P(X/C) ≥ P(C @ X) hold for all probabilities with P(C) > 0, then C ⊃ X entails C $ X and C @ X entails C & X.*

Proof. First let us prove that $C @ X$ entails $C \& X$. If this entailment did not go through, then both propositions $(C @ X) \& ¬(C \& X)$ and $C \& ¬X$ would be consistent. Since $(C @ X) \& ¬(C \& X)$ and $[(C @ X) \& ¬C] ∨ [(C @ X) \& ¬X \& C]$ are logically equivalent this leaves us with two cases to consider.

Case 1: $(C @ X) \& ¬C$ is consistent. Here we can assign probabilities in such a way that $P(C \& X) = 0$ and $P(C \& ¬X) = P((C @ X) \& ¬C) = 1/2$. It follows automatically that $P(C @ X) ≥ P((C @ X) \& ¬C) = 1/2 > 0 = P(X/C)$.

Case 2: $(C @ X) \& ¬C$ is inconsistent. Since $[(C @ X) \& ¬X \& C]$ must then be consistent, this allows one to assign probabilities in such a way that $P(C \& X) = 0$ and $P((C @ X) \& ¬X \& C) = 1$. It follows directly that $P(C @ X) = P((C @ X) \& ¬X \& C) > 0 = P(X/C)$.

In both these cases we have constructed probabilities with $P(C) > 0$ for which $P(X/C) ≥ P(C @ X)$ fails. Thus, if this inequality is to hold in general the @ operation must be logically stronger than conjunction.

To see why $ must be logically weaker than material implication it suffices to observe that $P(C $ X) ≥ P(X/C)$ can only hold for all probabilities with $P(C) > 0$ if $P(X/C) ≥ P(¬(C $ ¬X))$ also holds for all probabilities with $P(C) > 0$. We can apply the result just established with $¬(C $ ¬X)$ substituted for $C @ ¬X$ to conclude that $¬(C $ ¬X)$ must entail $C \& ¬X$, from which it follows that $C ⊃ X$ entails $C $ X$. This completes the proof of Theorem 6.2. ∎

Lemma 6.5. *Let $P(•//•)$ and $P*(•//•)$ be Réyni–Popper measures defined on $Ω × C$. Suppose that P*'s "anchor" probability is a Jeffrey shift of P's, so that $P*(•) = P(•//T) = p*P(•//D) + (1 - p*)P(•//¬D)$ for some condition $D ∈ C$ such that $1 > P(D) > 0$ and some $p* ∈ [0,1]$. Then, for any $C ∈ C$ such that both $P(C \& D)$ and $P(C \& ¬D)$ are positive, one will have*

(a) *$P*(X//C \& D) = P(X//C \& D)$ and $P*(X//C \& ¬D) = P(X//C \& ¬D)$*

222

(b) $P^*(X//C) = \dfrac{P^*(D)P(X \& C//D) + P^*(\neg D)P(X \& C//\neg D)}{P^*(D)P(C//D) + P^*(\neg D)P(C//\neg D)}$

$\qquad\qquad = P^*(D//C)P(X//C \& D) + P^*(\neg D//C)P(X//C \& \neg D)$

(c) $P^*(D//C) = \dfrac{P^*(D)P(C//D)}{P^*(D)P(C//D) + P^*(\neg D)P(C//\neg D)}$

$\qquad\qquad = \dfrac{P^*(D)[P(D//C)/P(D)]}{P^*(D)[P(D//C)/P(D)] + P^*(\neg D)[P(\neg D//C)/P(\neg D)]}$

Proof Sketch. For (a) use the fact that $P^*(\bullet//D) = P(\bullet//D)$ and $P^*(\bullet//\neg D) = P(\bullet//\neg D)$ to conclude that $P^*(X \& C//D)/P^*(C//D) = P(X \& C//D)/P(C//D)$. (b) is just a matter of noting that $P^*(C)$ must be positive since $P^*(C//D) = P(C//D) \geq P(C \& D) > 0$, and then applying Jeffrey's rule to the numerator and denominator of $P^*(X \& C)/P^*(C)$. The first identity in (c) is just (b) with D substituted for X. To get the second identity apply Bayes's Theorem to the first one to obtain $P(C//D) = P(D//C)[P(C)/P(D)]$ and $P(C//\neg D) = P(\neg D//C)[P(C)/P(\neg D)]$, and then clear the $P(C)$. This completes the proof. ∎

7

A Representation Theorem for Causal Decision Theory

Having come to grips with the concept of a conditional belief we now return to the problem of proving a representation theorem for causal decision theory. No decision theory is complete until it has been supplemented with a representation theorem that shows how its "global" requirement to maximize expected utility theory will be reflected at the "local" level as constraints on individual beliefs and desires. The main foundational shortcoming of causal decision theory has always been its lack of an adequate representation result. Evidential decision theory can be underwritten by Bolker's theorem and the generalization of it that was established at the end of Chapter 4. This seems to militate strongly in favor of the evidential approach. In this chapter I remove this apparent advantage by proving a Bolker-styled representation result for an abstract *conditional decision theory* whose two primitives are probability under a supposition and preference under a supposition. This theorem is, I believe, the most widely applicable and intuitively satisfying representation result yet attained. We will see that, with proper qualifications, it can be used as a common foundation for both causal decision theory and evidential decision theory. Its existence cements one of the basic theses of this work. It was claimed in Chapter 5 that evidential and causal decision theories should not be seen as offering competing theories of value, but as disagreeing about the epistemic perspective from which actions are to be evaluated. The fact that both theories can be underwritten by the same representation result shows that this is indeed the case.

7.1 THE WORK OF GIBBARD AND ARMENDT

The easiest way to prove a representation theorem for causal decision theory would be to co-opt some existing result by supplementing its axioms with constraints that capture utility as *efficacy value.*

Allan Gibbard has done this by using Savage's theorem, [1] and Brad Armendt has employed Peter Fishburn's conditional utility theory for the same purpose. [2] While both these results are interesting and important for what they tell us about causal decision theory's relationship to two of the standard formulations of expected utility theory, neither is ideal because the representation results of Savage and Fishburn are less than fully satisfactory.

Gibbard supplements the axioms that govern preferences in Savage's theory by two constraints on beliefs about subjunctive conditionals that together suffice to pick out an acceptable state partition relative to which causal utilities may be computed. Expressed in terms of a decision problem $D = (\Omega, O, S, A)$, these new axioms are[3]

Definiteness of Outcome: Let X be any proposition in Ω that the agent cares about (in the sense of not being indifferent between X and $\neg X$), and let A be any act in A. Then, either $[(A \,\Box\!\!\rightarrow X) \,\&\, S] .=. S$ or $[(A \,\Box\!\!\rightarrow \neg X) \,\&\, S] .=. S$ should hold for any state S in S.

Instrumental Act Independence: For any act A and any state S it should be the case that $[(A \,\Box\!\!\rightarrow S) \equiv S] .=. T$.

The first of these says that the agent must be certain about all the good or evil things that would accompany her acts when any given state obtains. The second says that she must be certain that the states in S are counterfactually independent of what she does. Under these conditions, Gibbard shows, the expected utility that Savage's axioms deliver for any act A will coincide with its efficacy value. This makes it possible for the causal decision theorist to use Savage's representation theorem as a foundation for causal decision theory subject to the proviso that Savage's axioms are only appropriately applied when the two stated conditions hold.

Gibbard's approach does, of course, alter the nature of Savage's theorem since it no longer characterizes expected utility maximization in terms of constraints on preferences alone. Some will see this as a disadvantage, but I think it quite appropriate. As I argued in connection with Bolker's theorem, it is wrong-headed to try to understand prudential reason by reducing the laws of rational belief to the laws of

[1] Gibbard (1984).

[2] Armendt (1986). Fishburn's theory is developed in Fishburn (1973).

[3] Gibbard does not actually use the comparative probability relation to express these requirements. Instead, he assumes a primitive notion of "knowing that" or "being certain" conditional on some proposition being true. So, where he speaks of the agent's *knowing X conditional on C*, I speak of her *being as confident in X & C as she is in C*. It should be clear that there is no substantive difference here.

rational desire since this both gives rise to unacceptable forms of pragmatism and forces theorists proving representation results to impose unduly strong structural constraints on preferences. Thus, I have no objection to Gibbard's talk of the agent's beliefs. I do, however, think it is unwise to employ Savage's representation theorem in this context because all the problems associated with it are thereby imported into causal decision theory. The main problems I have in mind are its use of "constant" acts and its inability to handle "small-world" decision making.

Armendt's representation theorem takes Fishburn's *conditional* decision theory as its starting point. The basic concept here is that of the utility of an action A on the hypothesis that some condition E obtains, written here as $U(A\|E)$. This quantity is governed by the (partition invariant) equation:

Fishburn's Equation. $U(A\|E) = \Sigma_S P(S/E)u(A\|S)$
$$= \Sigma_j P(E_j/E)U(A\|E_j)$$

where $\{E_1, E_2, \ldots E_n\}$ is any partition of E. For Armendt's purposes the most important thing about this formula is that it allows for a distinction between an act A's *unconditional* utility and its utility conditional on its own performance. These are given, respectively, by

$$U(A) = U(A\|\mathbf{T}) = \Sigma_S P(S)u(A\|S)$$
$$U(A\|A) = \Sigma_S P(S/A)u(A\|A \& S).$$

Notice how much $U(A)$ looks like Savage's Equation, and how much $U(A\|A)$ looks like Jeffrey's. Indeed, if one had $U(A\|S) = U(A\|A \& S) = U(A \& S)$, as one generally does *not*, then the top formula would be Savage's and the bottom one would be Jeffrey's

In a suggestion that bears similarities to Jeffrey's ratificationist proposal, Armendt argues that A's auspiciousness diverges from its efficacy value precisely in cases where unconditional preference for A differs from her preference for A conditional on itself. With respect to the act A_2 of refusing the extra \$1,000 in Newcomb's problem, for example, Armendt writes, "it is highly plausible that A_2 [given] A_2 is ranked below $A_2, \ldots$ Under the hypothesis that I [refuse], my preference for [refusing] is diminished, since worlds in which I [refuse] are worlds where an empty [bank account] is likely."[4] More generally, he claims that

---

[4] Armendt (1986, p. 10).

unconditional preference is perturbed by a conditional hypothesis to the extent that the hypothesis carries information which makes a difference to the agent's estimate of [a prospect] P's *value* or *utility*. But since to hypothesize is not to acquire news, preference is not perturbed by alterations in *degree of belief*... The hypothesis that a proposition is true or the hypothesis that it is false does not affect the [causal] utility that the agent attaches to it. But sometimes the hypothesis that the actual world is a P-world may carry information about states that the agent (believes are) correlated with P.[5]

Any difference between $U(A\|A)$ and $U(A)$ is, in Armendt's view, an indication that the state partition has been chosen incorrectly and thus that the value of $U(A)$ cannot be confidently used as a guide to action.

To isolate the right state partition (i.e., the one for which $U(A)$ is A's correct causal expected utility) Armendt imposes two additional constraints on the agent's preferences. Skipping over some of the technicalities, his first idea is that the elements of an appropriate state partition will "screen off" differences in value between A and A conditional on itself. In our terms this can be expressed as follows:

Value Screening. For each state S and act A, the decision maker should be indifferent between A given S and A given A & S, so that $A\|S \approx A\|(A \& S)$.

When this holds, $U(A) = \Sigma_S P(S)u(A\|A \& S)$ and $U(A\|A) = \Sigma_S P(S/A)u(A\|A \& S)$. These equations stand to one another as Savage's and Jeffrey's do. As Armendt goes on to observe, the conditional utilities $U(A\|A \& S)$ can be eliminated in favor of unconditional news values $V(A \& S)$ as long as there exists at least one partition of "consequences" $\boldsymbol{C}$ such that for any proposition $C \in \boldsymbol{C}$ the agent's utility for A conditional on $(A \& C \& S)$ always equals her unconditional utility for $(A \& C \& S)$.

Existence of Consequences: There is a partition of propositions $\boldsymbol{C}$ such that $A\|(A \& C \& S) \approx (A \& C \& S)$ for all states S and acts A and all $C \in \boldsymbol{C}$.

If this is the case then $U(A\|A \& S) = \Sigma_C P(C/A \& S)U(A\|A \& C \& S)$ and the equation for unconditional utility becomes

$$U(A) = \Sigma_S P(S)[\Sigma_C P(C/A \& S)U(A \& C \& S)]$$

This is a version of the $\boldsymbol{K}$ expectation formula for causal decision theory discussed in Chapter 5. Armendt takes this similarity of

[5] Armendt (1986, p. 10).

form to show that his two constraints capture the appropriate notion of a K partition for use in calculating efficacy values. If this is right, then adding them to the axioms for Fishburn's conditional utility theory will provide a representation theorem for causal decision theory.

This is a nice idea. Armendt is right to think that one should look to a theory of conditional expected utility to find a foundation for causal decision theory, and his characterization of K partitions is quite suggestive. Still, I am not sure that he has made the case for Value Screening as a hallmark of the partitions that should be used to compute causal utilities. It is hard to see how it captures the intuitive notion of an element of K as "a complete description of the ways in which things the agent cares about might depend on what she does," or, indeed, to see where causality comes into the picture at all. How do the indifferences $A\|S \approx A\|(A \ \& \ S)$ and $A\|(A \ \& \ C \ \& \ S) \approx (A \ \& \ C \ \& \ S)$ reflect facts about the causal connection between A and C in the presence of S? I am willing to be open-minded here – it may be that Armendt really has characterized the role that K partitions play in rational preference rankings – but the case needs to be more clearly made.

Even if it is, however, Armendt's result would still not provide a sound foundation for casual decision theory. The problem is not his work, but Fishburn's. While Fishburn's equation is partition-invariant (for a given S), and while his representation result does not employ constant acts, it does assume that *mitigators* exist. Recall from Chapter 3 that a mitigator is an act that is able to offset whatever desirable or undesirable things might occur, for example, an act whose performance can make a person indifferent between the prospect of an asteroid destroying all life on earth in the next five minutes and the prospect of peace and prosperity for a millennium. Fishburn's theory includes an axiom which says, in effect, that for any two events E and F there is an action A such that the agent prefers A on the condition that E to A on the condition that F. Let $E = $ "An asteroid hits the earth in the next five minutes and destroys all life" and $F = $ "There is peace, prosperity, and happiness everywhere on earth for a thousand years." Try to think of an appropriate A, but don't try too hard because there is none. As we saw in Chapter 3, representation theorems that make use of constant acts or mitigators are to be avoided because these are the sorts of structure requirements that cannot be explained away by construing them as extendibility conditions. For this reason we cannot use Fishburn's theory as a basis for causal decision theory.

228

The only live option here, in my view, is the extended version of Bolker's theorem that was proved Chapter 4. As we saw, this result is particularly appealing from the formal point of view. It requires only two nonnecessary structure axioms – a completeness axiom and a non-atomicity condition – both of which are plausible as canons of rationality when viewed as extendibility requirements. Moreover, unlike Fishburn or Savage, Bolker does not make essential use of preferences over prospects that a reasonable agent might regard as impossible, such as constant acts or mitigators. In the next section I will show how to prove a general representation theorem for causal decision theory on the basis of Bolker's theorem. I will do this by proving a general representation result for *conditional* expected utility theory, and then showing how both causal and evidential decision theory can be seen as instances of it.

7.2 A STATEMENT OF THE THEOREM

The representation result we are after assumes an agent facing a decision $\mathbf{D} = (\Omega, \mathbf{O}, \mathbf{S}, \mathbf{A})$ whose beliefs are described by a *conditional likelihood ranking* $(\|.>.\|, \|.\geq.\|)$ defined relative to a set of conditions $\mathbf{C}$ that contains all the acts in $\mathbf{A}$ (and perhaps other propositions), and whose desires are described by a *conditional* preference ranking $(\|>\|, \|\geq\|)$ also defined relative to $\mathbf{C}$. Since I will be assuming that both these rankings are complete for the purposes of this proof I will simply use $\|.\geq.\|$ for $(\|.>.\|, \|.\geq.\|)$ and $\|\geq\|$ for $(\|>\|, \|\geq\|)$. This greatly simplifies the presentation, but the reader should remain aware that this completeness requirement should ultimately be dispensed with, so that the full story is told in terms of the incomplete rankings $(\|.>.\|, \|.\geq.\|)$ and $(\|>\|, \|\geq\|)$.

We seek a set of axiomatic constraints on $\|.\geq.\|$ and $\|\geq\|$ that suffice for the existence of a pair of functions $P(\bullet\|\bullet)$ and $V(\bullet\|\bullet)$ defined on $\Omega \times \mathbf{C}$ such that

7.1a. $P(\bullet\|\bullet)$ is a supposition function; that is, for each $C \in \mathbf{C}$ one has

SUP$_1$ (*Coherence*): $P(\bullet\|C)$ is a probability on Ω.
SUP$_2$ (*Certainty*): $P(C\|C) = 1$.
SUP$_3$ (*Regularity*): $P(X\|C) \geq P(X \,\&\, C)$ when $C \in \mathbf{C}$.

7.1b. For each $C \in \mathbf{C}$, $V(\bullet\|C)$ gives expected news values computed relative to $P(\bullet\|C)$, so that for all $X \in \Omega$

$$V(X\|C) = \Sigma_W \frac{P(W \,\&\, A\|C)}{P(A\|C)} u(W)$$

229

where u is a function that assigns an unconditional utility $u(W)$ to each atomic proposition W in Ω.[6]

7.1c. $P(\bullet\|\bullet)$ ordinally represents $\|.\geq.\|$, and $V(\bullet\|\bullet)$ ordinally represents $\|\geq\|$.

7.1d. $P(\bullet\|\bullet)$ is unique, and $V(\bullet\|\bullet)$ (and hence u) is unique up to the arbitrary choice of a unit and a zero point relative to which utility is measured.

Once we have a theorem like this it is straightforward to use it to underwrite either causal decision theory or evidential decision theory. We merely need to impose constraints on the decision maker's conditional likelihood ranking that are strong enough to determine that it represents her beliefs under the right sorts of suppositions. So, if we are interested in a representation result for evidential decision theory $\|.\geq.\|$ must satisfy

MOF (comparative version). If $C//D.>.\neg\mathbf{T}//D$, then $X//(C\ \&\ D)$ $.\geq.\ Y//(C\ \&\ D)$ if and only if $(C\ \&\ X)//D.\geq.(C\ \&\ Y)//D$.

In the presence of the other axioms this uniquely picks out the standard conditional probability $P(\bullet/\bullet)$ as the value of $P(\bullet\|\bullet)$ relative to $\mathbf{C} = \{C \in \Omega: C.>.\ C\ \&\ \neg C\}$, and, more generally, it forces $P(\bullet\|\bullet)$ to be a Réyni–Popper measure.

The causal decision theorist, on the other hand, can impose whatever conditions she thinks necessary to have $\|.>.\|$ capture the decision maker's causal beliefs. As we have seen, one natural constraint here is

SUB (Comparative Version). If$(C \to X)\backslash\backslash\mathbf{T}.\geq.\neg(D \to \neg Y)\backslash\backslash\mathbf{T}$, then $X\backslash\backslash C.\geq.Y\backslash\backslash D$

Or, when Conditional Excluded Middle holds,

$(C \to X)\backslash\backslash\mathbf{T}.\geq.(D \to Y)\backslash\backslash\mathbf{T}$ if and only if $X\backslash\backslash C.\geq.Y\backslash\backslash D$

Other, stronger requirements could be imposed on the basis of the view that one takes about the proper analysis of causal judgments. Whatever these requirements are, however, they would be always imposed on top of the ones already given and can therefore be neglected in the present context.

Now it might seem as if there is not much being offered to the evidential decision theorist here. They do, after all, already have

---

[6] Recall that atomic outcomes are act/state conjunctions $A\ \&\ S$ where $A \in \mathbf{A}$ and $S \in \mathbf{S}$.

Bolker's theorem as the foundation for their theory and this "generalization" really does not do much since "conditional" utilities reduce to unconditional utilities of conjunctions when the supposition function is indicative; that is, $V(X \| C) = V(X \& C)$ when $P(\bullet \| C) = P(\bullet / C)$. So, the appearance of a "unified" foundation for causal and evidential decision theory might be illusory. I am happy to grant that there is nothing new in this result as it applies to evidential decision theory as standardly formulated. Its advantage is that it allows evidential decision theorists to extend their theory to allow for news values defined in terms of Réyni–Popper functions. There is a good reason for them to want to do so.

As a number of critics have noted,[7] Jeffrey's theory seems to lead to absurd results in cases where an agent is certain about what she will do. When $P(A) = 1$ Jeffrey's Equation sets $V(A) = V(A \vee \neg A)$ and leaves $V(\neg A)$ undefined (though Jeffrey conventionally sets it to 0). The problem with this is that it makes it appear as if "awareness of [one's] preference for [one's] top-ranked option over $A \vee \neg A$ reduces preference to indifference."[8] It thus becomes impossible for one to speak sensibly about the evidentiary value of acts one has irrevocably decided to perform. Jeffrey has responded to this difficulty by (i) distinguishing the utility, V, that represents the agent's desires *before* she makes up her mind from the utility, V_A, that represents her desires after she is sure she will do A, and (ii) pointing out that $V(A \vee \neg A)$ and $V_A(A \vee \neg A)$ may differ.[9] The agent, in other words, need not be portrayed as being indifferent between the "*status quo*" before she becomes certain of A and the *status quo* afterward. With this distinction in place we can say that she sees herself as better off for having done A just in case $V_A(A \vee \neg A) > V(A \vee \neg A)$.

While this is right as far as it goes, it leaves a crucial issue unresolved. On Jeffrey's proposal a person who is certain she will perform A must still assign the same news value to every act incompatible with A, and this makes it impossible for her to compare news values of acts she is sure she will *not* perform. The most she can say is that as far

---

[7] See, for example, Sphon (1977, p. 113).

[8] Jeffrey (1977, p. 136).

[9] Jeffrey expresses this point by saying that the contradictory proposition need not appear at the same place in the agent's preperformance and postperformance preference rankings. Instead of saying that $V(A \vee \neg A)$ and $V_A(A \vee \neg A)$ may differ, he says that $V(A \vee \neg A) - V(A \& \neg A)$ and $V_A(A \vee \neg A) - V_A(A \& \neg A)$ may differ where it is understood that he is keeping the news value of $A \vee \neg A$ set at 0. The more intuitive way of making the point, it seems to me, is to keep the news-value of $A \& \neg A$ fixed, say at 0, and to let that of $A \vee \neg A$ vary, depending on what the decision maker does. This coheres better with the idea, expressed in the previous chapter, that the goal of action is to produce the best postaction *status quo*.

as auspiciousness goes, A is better than the alternatives. This is not an ideal result. Even someone who is sure she will do A should still be able to evaluate her other alternatives and say things like "The most auspicious option among the ones I did not choose was B" where this not merely a statement about her *past* evaluations of acts, but an expression of her *current* view of the situation. Comparisons like this, after all, often figure into our justifications of acts; for example, I might be sure I am going to do (or did) A *because* I recognize that A is better than B and that B is better than C, where B and C are acts I know I will not (or did not) perform. Jeffrey's approach lets me say the first thing but not the second. If I am going to be allowed to make discriminations in evidential expected utility among acts I am sure not to perform, then evidential decision theory's basic equation must be rewritten so that $V(B)$ can be well defined even when $P(B) = 0$.

The best way to do this is by substituting a Réyni–Popper measure for the ordinary conditional probability in Jeffrey's equation so that it becomes $V(B) = \Sigma_S P(S//B)u(S \ \& \ B)$. Since $P(S//B)$ can be well defined and positive even when $P(B) = 0$, this allows a decision maker to draw distinctions in news value among actions that she is quite sure she will not perform.[10] A person who has irrevocably decided to take the extra thousand dollars in the Newcomb problem can, for example, still make sense of the idea that refusing it would be a more auspicious act, not just from the perspective of her predecision beliefs but from her *current* epistemic position. A fully adequate account of the auspiciousness of acts will thus need to traffic in Réyni–Popper measures. And, if this is so, then evidential decision theorists are going to need a new representation theorem because Bolker's only provides an expected utility representation for propositions that are nonnull relative to the decision maker's preference ranking. This new theorem will need to make sense of true conditional news values that are not mere unconditional news values of conjunctions. The result they will need is the one we are about to prove.

---

[10] In his unpublished work Frank Doring has independently suggested that Jeffrey's theory needs to be formulated in terms of Réyni–Popper measures. While Doring is right about this point his main motivation for accepting it has to do with finding a way of making sense of "if only I had done A" evaluations of actions. Since these evaluations have a subjunctive character I do not regard them as being appropriately captured by Réyni–Popper measures. To reemphasize, the Réyni–Popper measures are *not* suited to capturing subjunctive beliefs. Thus, a news-value for an act B that an agent is sure she will not perform need not be the same as the act's efficacy value (as the next sentence in the text indicates).

My proof strategy is going to be one of divide and conquer. For each condition C in $\boldsymbol{C}$, define the *C-section* as the unconditional likelihood/preference ranking pair defined by

$$X .\geq^C. Y \text{ if and only if } X\|C .\geq. Y\|C$$

$$X .\underset{\cdot}{\geq}^C. Y \text{ if and only if } X\|C \geq Y\|C$$

Since the function defined in 7.1c is a news value for every C, it makes sense to impose the Jeffrey/Bolker axioms of Chapter 4 on each C-section individually. This produces a set of *sectional representations* SR = {(P($\bullet\|C$), V($\bullet\|C$)): $C \in \boldsymbol{C}$} where each (P($\bullet\|C$), V($\bullet\|C$)) pair is a Bolker-style representation of its associated C-section. The challenge will be to stitch these sectional representations together in the right way to get a full joint representation for $\|.\geq.\|$ and $\|\geq\|$.

It requires three axioms (really two axioms and a general principle) to ensure that the requisite system of sectional representations SR will exist. The first describes the behavior of propositions that the decision maker regards as certainly true.

Axiom₁ (*Certainty*). *If $C .=. T$, then $.\geq^C.$ is identical to $.\geq^T.$, and $\underset{\cdot}{\geq}^C$ is identical to $\underset{\cdot}{\geq}^T$.*

This says that the supposition of propositions that an agent regards as certain should not alter her beliefs or desires. When she supposes that some proposition C is true, the agent adopts a new set of beliefs $.\geq^C.$ that makes C certain and does as little damage as possible to her prior opinions $.\geq^T.$. Axiom₁ merely says that if she already takes C to be certain, then the new belief system that approximates the old one most closely is the old system itself. The requirement that $\underset{\cdot}{\geq}^C$ should not change is a consequence of the fact that supposition is an *epistemic* operation that affects belief directly and alters desires only through the mediation of beliefs. Supposition, in other words, never changes the decision maker's *basic* desires.

The second rationality requirement for C-sections is

Axiom₂ (*Conditional Certainty*). $C .=^C. T$.

This says that belief given C should be genuinely based on the supposition of C's truth. In the presence of the other axioms, this implies that things that happen when C is false are irrelevant to beliefs and

233

desires conditional on C. Thus, the agent will always regard X as precisely as likely and desirable as $X \& C$ when she supposes C.

Our final constraint on C-sections requires beliefs and desires conditional on C to obey the same laws of rationality that apply to unconditional beliefs and desires:

Axiom$_3$ (*Conditional Rationality*). *C-sections must obey the same laws of rationality that apply to unconditional likelihood and preference rankings.*

This demands that $.\geq^C.$ and $\geq^C$ be evaluated with regard to rationality in the same way that any other unconditional likelihood/preference ranking pair would be. It says, in other words, that a person should be bound by the same laws of rationality when she supposes some hypothesis to be true as when she supposes nothing at all. I have been careful to state this principle in a way that does not presuppose any specific analysis of rationality for unconditional beliefs or desires because I believe that its validity should be affirmed independently of any disagreements there may be about the particulars of such an analysis. This is a view with which I think most decision theorists would agree.[11]

That having been said, I mean to defend a version of Axiom$_3$ that does take a stand on the nature of rationality for unconditional beliefs and desires. As I argued at the end of Chapter 5, I think all value is news value. Thus, I will require each C-section to satisfy the axioms employed in the version of Bolker's theorem established at the end of Chapter 4. Accordingly, my official version of Axiom$_3$ will be

Axiom$_3$. *For each $C \in \mathbf{C}$, $.\geq^C.$ should satisfy the laws of comparative probability CP_1–*CP_8, $\geq^C$ must obey the Jeffrey/Bolker axioms EDT_1–*EDT_9, and $.\geq^C.$ and $\geq^C$ should jointly obey Coherence.*

In other words, each belief/desire pair $.\geq^C.$ and $\geq^C$ must be *EDT-coherent* in the terminology of Chapter 4. I suspect that at this point even the most conscientious readers will have forgotten what this means. The only two things about EDT-coherence that matter at the moment are that (i) it forces the base algebra Ω to be atomless with respect to $.\geq^C.$, and (ii) it ensures the existence of a joint probability/ news-value representation $(P(\bullet\|C), V(\bullet\|C))$ of $.\geq^C.$ and $\geq^C$ in which

[11] See, for example, Savage (1954/1972, p. 78).

$P(\bullet\|C)$ is unique and $V(\bullet\|C)$ is unique up to the arbitrary choice of a zero point and a unit for measuring utility.

I take $Axiom_1$–$Axiom_3$ to be the fundamental laws governing rational belief and desire conditional on a single hypothesis C. They take us some way toward a representation result for conditional likelihood and preference rankings.

Lemma 7.2 (*Existence of Sectional Representations*). *If* $.\geq^C.$ *and* $\geq^C$ *satisfy* $Axiom_1$–$Axiom_3$ *for every* $C \in \boldsymbol{C}$, *then there is a sectional representation*

$$SR = \{(P(\bullet\|C), V(\bullet\|C)) : C \in \boldsymbol{C}\}$$

in which

i. $P(\bullet\|C)$ *is a countably additive probability on* Ω *with* $P(C\|C) = 1$.
ii. $V(X\|C) = \Sigma_W [P(W \& X\|C)/P(X\|C)]V(W\|C)$ *when* $P(X\|C) > 0$.
iii. $P(\bullet\|C)$ *represents* $.\geq^C..$
iv. $V(\bullet\|C)$ *represents* $\geq^C$.

Moreover, any other sectional representation for $.\geq^C.$ *and* $\geq^C$ *will have the form*

$$SR* = \{(P(\bullet\|C), a_C V(\bullet\|C) + b_C) : C \in \boldsymbol{C}\}$$

for a_C *and* b_C *real numbers (dependent on C) with* $a_C > 0$.

This lemma codifies what we can say about the rationality of a system of conditional beliefs and desires when we restrict our attention to beliefs and desires under the supposition that a single condition is true.

It does not, however, tell us anything about "mixed" beliefs and desires in which a decision maker judges that X is more likely or more desirable given C than Y is given D. Since "mixed" beliefs and desires of this type are important to the evaluation of actions, we need to extend Lemma 7.2 to cover this case. It does not do so automatically. While any full representation for $\|.\geq.\|$ and $\|\geq\|$ is a sectional representation, the converse is not the case. In fact, nothing we have said to this point guarantees that *any* of these sectional representations is a full representation. It is, for example, consistent with $Axiom_1$–$Axiom_3$ that $\|\geq\|$ is intransitive (even though all of its individual sections are transitive). We must introduce additional axioms if we want to establish the existence of a full representation of the desired type.

There has been a great deal of work done on the representation of ordinary conditional probability functions.[12] The result we need, however, is slightly more general than any that can be found in the literature because we want it to be possible for $\|.>.\|$ to describe nonevidential forms of belief revision like imaging. The axiom required to obtain the desired representation is

Axiom$_4$. *For and C, D, E ∈ $\boldsymbol{C}$ and W, X, Y, Z ∈ Ω, $\|.\geq.\|$ must satisfy*

- *Normalization: $C\|C .=. D\|D$ and $\neg C\|C .=. \neg D\|D$.*
- *Transitivity: If $X\|C .\geq. Y\|D$ and $Y\|D .\geq. Z\|E$, then $X\|C .\geq. Z\|E$.*
- *Connectedness: Either $X\|C .\geq. Y\|D$ or $X\|C .\leq. Y\|D$.*
- *Dominance: If W and X are logically incompatible, and if Y and Z are also incompatible, then $W\|C .\geq. Y\|D$ and $X\|C .\geq. X\|D$ only if $(W \vee X)\|C .\geq. (Y \vee Z)\|D$.*
- *Regularity: $X\|C .\geq. (X \& C)\|\boldsymbol{T}$.*
- *Solvability: If $X\|C .\geq. Y\|D$, then there exists $X^* \in \Omega$ such that $(X^* \& X)\|C .=. Y\|D$.*

The only one of these that is not self-explanatory (by this point in this book) is Solvability. It is an Archimedean axiom that rules out infinitesimal probabilities. Since $.\geq^C.$ is atomless (by Axiom$_3$), it follows that, for any proposition $X \in \Omega$ that is nonnull with respect to $.\geq^C.$ the set of numbers $\{P(X \& X^*\|C): X^* \in \Omega\}$ will contain every value in the interval from 0 to $P(X\|C)$. Therefore, if $P(X\|C) > P(Y\|D)$ and both these probabilities are real numbers, then $P(X^*\|C) = P(Y\|D)$ should hold for some X^*.

Using Axiom$_4$ one can establish

Theorem 7.3 (*Existence of Probability Representations for Suppositions*). *If $\|.\geq.\|$ and $\|\geq\|$ satisfy Axiom$_1$–Axiom$_3$ and if SR = $\{(P(\bullet\|C), V(\bullet\|C)): C \in \boldsymbol{C}\}$ is any sectional representation for $(.\geq^C., \geq^C)$, then Axiom$_4$ is necessary and sufficient for*

7.1a SUP$_1$: $P(\bullet\|C)$ is a countably additive probability on Ω.
   SUP$_2$: $P(C\|C) = 1$.
   SUP$_3$: $P(X\|C) \geq P(X \& C)$ when $C \in \boldsymbol{C}$.

7.1c $P(\bullet\|\bullet)$ ordinally represents $\|.>.\|$.

7.1d $P(\bullet\|\bullet)$ is the only function for which 7.1a and 7.1c hold.

---

[12] See, for example, Fine (1973, p. 29).

(Interested readers can find the proof of Theorem 7.3 in the Appendix of this chapter.) The most important thing to notice about this result is its uniqueness clause, which ensures that every sectional representation for $\|.\geq.\|$ and $\|\geq\|$ must involve *the same* supposition function. This is a consequence of the Ordinal Uniqueness Theorem that was established in Chapter 4. The uniqueness of $P(\bullet\|\bullet)$ turns out to be crucial in what follows.

To obtain a joint representation for $\|.\geq.\|$ and $\|\geq\|$, we must introduce supplementary axioms to clarify the nature of conditional preference and its relationship to belief. Here are the "desire specific" conditions that pertain to the preference ranking $\|\geq\|$:

Axiom₅. *For and $C, D, E \in \mathbf{C}$ and $W, X, Y, Z \in \Omega$, $\|.\geq.\|$ must satisfy*

- *Transitivity: If $X\|C \geq Y\|D$ and $Y\|D \geq Z\|E$, then $X\|C \geq Z\|E$.*
- *Connectedness: Either $X\|C \geq Y\|D$ or $X\|C \leq Y\|D$.*
- *Invariance of Basic Desires: If W is an atom of Ω, then $W\|C \approx W\|D$ for all $C, D \in \mathbf{C}$.*
- *Solvability: Let X and Y be nonnull relative to C, so that $X\|C .>.$ $\neg C\|C$ and $Y\|C .>. \neg C\|C$. For any $Z \in \Omega$ and $D \in \mathbf{C}$ such that $X\|C \geq Z\|D \geq Y\|C$ there exists $X^* \in \Omega$ with $X^*\|C \approx Z\|D$.*
- *Topological Separability: If $\mathbf{D}$ is a subset of conditions in $\mathbf{C}$ such that, for all $C, D \in \mathbf{C}$, one has either $X\|C > Y\|D$ for all $X, Y \in \Omega$ or $X\|C < Y\|D$ for all $X, Y \in \Omega$, then $\mathbf{D}$ must be countable.*

The first two principles require conditional preferences to be transitive and connected. The third expresses the idea that supposition does not affect basic desires. The role of the solvability condition here is the same as it was in Axiom₄. For a given probability/news value pair $P(\bullet\|C)$ and $V(\bullet\|C)$, define the *essential range* of $V(\bullet\|C)$ as the collection of numbers $\{V(X\|C): P(X\|C) > 0\}$. When $\|.\geq.\|$ and $\|\geq\|$ satisfy Axiom₃ the essential range of any utility $V(\bullet\|C)$ will be an interval on the real line. (See Fact 4 in the Appendix at the end of this chapter.) Hence, if $X\|C \geq Z\|D \geq Y\|C$ holds when X and Y are nonnull with respect to $\|.\geq.\|$, then in any real-valued representation $V(\bullet\|\bullet)$ of $\|\geq\|$ it must be the case that $V(X^*\|C) = V(Z\|D)$ for some X^*. If this were not so, $V(Z\|D)$ could not be any real number. The separability condition ensures that the representing function $V(\bullet\|\bullet)$ can fit into the real line. To get a sense of its meaning, note that if its antecedent holds and if $P(\bullet\|\bullet)$ and $V(\bullet\|\bullet)$ are any representations of $\|.\geq.\|$ and $\|\geq\|$, then the essential ranges of all the functions $V(\bullet\|D)$, $D \in \mathbf{D}$, will form a family of *disjoint* intervals on the real line each of which has a nonempty

237

interior. Since there can be at most countably many such intervals, the set $\boldsymbol{D}$ must be countable.

It will be convenient to introduce some terminology in connection with these last two requirements that will set up our next axiom. Say that the sections $(.\geq^{C}., \, \geq^{C})$ and $(.\geq^{D}., \, \geq^{D})$ are *linked* when there are nonnull pairs $X_1\|C$, $X_2\|C$, and $Y\|D$ such that $X_1\|C > Y\|D > X_2\|C$. Given Solvability and Separability, this means that, in any representation of $(\|.\geq.\|, \|\geq\|)$, the essential ranges of $V(\bullet\|C)$ and $V(\bullet\|D)$ will have a nonempty intersection that contains an open interval of numbers. A *chain* is a countable sequence of sections $(.\geq^{C1}., \, \geq^{C1})$, $(.\geq^{C2}., \, \geq^{C2})$, $(.\geq^{C3}., \, \geq^{C3})$, . . . , such that $(.\geq^{C1}., \, \geq^{C1})$ is linked to $(.\geq^{C2}., \, \geq^{C2})$, $(.\geq^{C2}., \, \geq^{C2})$ is linked to $(.\geq^{C3}., \, \geq^{C3})$, $(.\geq^{C3}., \, \geq^{C3})$ is linked to $(.\geq^{C4}., \, \geq^{C4})$, and so on. Two sections that appear in the same chain are *fettered*.

Since we are aiming for a representation in which desires are represented by real-valued expected utilities defined over a set of atomic propositions it makes sense to demand that any two sections in $(\|.\geq.\|, \|\geq\|)$ should be fettered. This is not strictly required by the existence of the desired representation, but any representation for which it fails will be very, very odd.[13] I propose to rule them out by fiat:

Axiom$_6$. *Any two sections of $(\|.\geq.\|, \|\geq\|)$ are fettered.*

It turns out (as a result of Lemma 7.6) that the linking relation is symmetric, so this axiom makes $(\|.\geq.\|, \|\geq\|)$ into one big chain.

Our next axiom is a generalization of the Coherence principle of Chapter 4. It serves as the fundamental principle of rationality connecting $\|.\geq.\|$ and $\|\geq\|$. To introduce it we need yet another a piece of terminology.

Definition. *A test configuration is a four-tuple $(X_1\|C, \, X_2\|C, \, Y_1\|D, \, Y_2\|D)$ for which all of the following hold*

- *X_1 and X_2 are mutually incompatible propositions such that $X_1\| > (X_1 \vee X_2)\|C > X_2\|C$.*
- *Y_1 and Y_2 are mutually incompatible propositions such that $Y_1\|D > (Y_1 \vee Y_2)\|D > Y_2\|D$.*
- *$X_1\|C \approx Y_1\|D$ and $X_2\|C \approx Y_2\|D$.*

---

[13] For example, there must be real numbers $x > y$ and a proposition X such that (i) the utility $u(W)$ of every atom W that entails X falls above x, (ii) the utility of every atom that entails and $\neg X$ falls below y, and (iii) there is no proposition Z such that $P(Z\|X)$ and $P(Z\|\neg X)$ are both nonzero.

A test configuration gives us a way of comparing X_1's conditional probability given $X_1 \vee X_2$ in the C-section with Y_1's conditional probability given $Y_1 \vee Y_2$ in the D-section. In general, if $(X_1\|C, X_2\|C, Y_1\|D, Y_2\|D)$ is a test condition, and if $P(\bullet\|\bullet)$ and $V(\bullet\|\bullet)$ is a representation of $\|.\geq.\|$ and $\|\geq\|$, then $(X_1 \vee X_2)\|C \geq (Y_1 \vee Y_2)\|D$ will hold if and only if $P(X_1\|C)/P(X_1 \vee X_2\|C) \geq P(Y_1\|D)/P(Y_1 \vee Y_2\|D)$ *or, equivalently, if and only if* $P(X_1\|C)/P(X_2\|C) \geq P(Y_1\|D)/P(Y_2\|D)$. Our next axiom requires the relationships among probabilities determined in this way by the agent's preferences to cohere with her beliefs.

Axiom$_7$ (*Generalized Coherence*). *Let $(X_1\|C, X_2\|C, Y_1\|D, Y_2\|D)$ be a test configuration. Then,*

- *If $X_2\|C .=. Y_2\|D$, then $(X_1 \vee X_2)\|C \geq (Y_1 \vee Y_2)\|D$ if and only if $X_1\|D .\geq. Y_1\|D$.*
- *If $X_1\|C .=. Y_1\|D$, then $(X_1 \vee X_2)\|C \geq (Y_1 \vee Y_2)\|D$ if and only if $X_2\|D .\leq. Y_2\|D$.*
- *If $(X_1 \vee X_2)\|C .=. (Y_1 \vee Y_2)\|D$, then $(X_1 \vee X_2)\|C \geq (Y_1 \vee Y_2)\|D$ if and only if $X_1\|D .\geq. Y_1\|D$.*

All these clauses are different ways of saying that an agent's preferences only force $P(X_1\|C)/P(X_2\|C)$ to be greater than $P(Y_1\|D)/P(Y_2\|D)$ when her beliefs do not force the opposite inequality to be true.

7.5 CONSTRUCTING THE REPRESENTATION

We now have the resources we need to construct a representation for conditional utility theory. Here is the result:

Theorem 7.4 (*Existence of Conditional Utility Representations*). *If $\|.\geq.\|$ and $\|\geq\|$ satisfy Axiom$_1$–Axiom$_7$, then there is a pair of functions $P(\bullet\|\bullet)$ and $V(\bullet\|\bullet)$ defined on $\Omega \times \boldsymbol{C}$ such that*

7.1a $P(\bullet\|\bullet)$ is a supposition function.
7.1b For each $C \in \boldsymbol{C}$ and $X \in \Omega$,

$$V(X\|C) = \Sigma_W \frac{P(W \,\&\, A\|C)}{P(A\|C)} u(W)$$

for some function u that assigns an unconditional utility u(W) to each atomic proposition $W \in \Omega$.

7.1c P(•||•) represents ||.≥.||, and V(•||•) represents ||≥||.

7.1d P(•||•) is unique, and V(•||•) (and hence u) is unique up to the arbitrary choice of a unit and a zero point relative to which utility is measured.

I will merely sketch the argument of Theorem 7.4 here, leaving the proofs of the more difficult lemmas to the last section.

To begin, suppose that (||.≥.||, ||≥||) satisfies the axioms. Lemma 7.2 entails the existence of a sectional representation SR = {(P(•||C), V(•||C)): C ∈ **C**} for (||.≥.||, ||≥||), and Theorem 7.3 ensures that its suppositional probability P(•||•) is unique. It also follows from Lemma 7.2 that any other sectional representation for (||.≥.||, ||≥||) will have the form SR* = {(P(•||C), a_CV(•||C) + b_C): C ∈ **C**} where $a_C > 0$ and b_C are real constants that depend on C. The key to proving Theorem 7.4 lies in finding the right constants to make SR* a full representation for (||.≥.||, ||≥||).

We can start by asking how to construct a joint representation for two linked sections $(.\geq^C., \geq^C)$ and $(.\geq^D., \geq^D)$. To do this we must find a_C, b_C, a_D and b_D, with a_C and a_D positive, such that

7.5. $X\|C \geq Y\|D$ if and only if $a_C V(X\|C) + b_C \geq a_D V(Y\|D) + b_D$

whenever $X\|C$ and $Y\|D$ are nonnull. To find these numbers we rely on the following important fact about linked sections:

Lemma 7.6. *If $(.\geq^C., \geq^C)$ and $(.\geq^D., \geq^D)$ are linked, then there exists at least one test configuration of the form $(X_1\|C, X_2\|C, Y_1\|D, Y_2\|D)$.*

Since $X_1\|C \approx Y_1\|D$. $X_2\|C \approx Y_2\|D$ holds in any such test configuration we know that the desired values of $a_C, b_C, a_D,$ and b_D must be such that $a_C V(X_1\|C) + b_C = a_D V(Y_1\|D) + b_D$ and $a_C V(X_2\|C) + b_C = a_D V(Y_2\|D) + b_D$. This forces it to be the case that

7.7. $a_D = a_C[V(X_1\|C) - V(X_2\|C)]/[V(Y_1\|D) - V(Y_2\|D)]$
$b_D = a_C V(X_1\|C) - a_D V(Y_1\|D) + b_C$

These turn out to be the crucial relationships involved in obtaining a joint representation for a pair of linked sections. Their importance is due to

Lemma 7.8. *If $(.\geq^C., \geq^C)$ and $(.\geq^D., \geq^D)$ are linked, and if $a_C > 0$ and b_C are chosen arbitrarily and a_D and b_D are defined as in 7.7, then for all propositions $X, Y \in \Omega$ such that neither $X\|C$ nor $Y\|D$ is null one has*

7.5. $X\|C \geq Y\|D$ if and only if $a_C V(X\|C) + b_C \geq a_D V(Y\|D) + b_D$

Moreover, the values of a_D and b_D given by 7.7 are the only ones for which 7.5 holds.

Lemma 7.8 is the heart of my representation theorem. Its proof can be found at the end of this chapter.

The method we used to jointly represent $(.\geq^C., \geq^C)$ and $(.\geq^D., \geq^D)$ can be extended to a chain $(.\geq^{C1}., \geq^{C1})$, $(.\geq^{C2}., \geq^{C2})$, $(.\geq^{C3}., \geq^{C3})$, ... Since $(.\geq^{Cj}., \geq^{Cj})$ and $(.\geq^{Cj+1}., \geq^{Cj+1})$ are linked for each j, Lemma 7.6 gives us a test configuration $(X_j\|C_j, X_j*\|C_j, Y_j\|C_{j+1}, Y_j*\|C_{j+1})$ for each j. Choose an index k at random, and fix $a_k > 0$ and b_k arbitrarily. Lemma 7.8 then lets us use 7.7 to recursively define a *unique* series of pairs of constants (a_1, b_1), (a_2, b_2), ..., (a_k, b_k), (a_{k+1}, b_{k+1}), ... such that

$$X\|C_j \geq Y\|C_{j+1} \text{ if and only if } a_j V(X\|C_j) + b_j \geq a_{j+1} V(Y\|C_{j+1}) + b_{j+1}$$

holds for all j and all $X, Y \in \Omega$. Let's call the construction that results in the sequence of pairs (a_1, b_1), (a_2, b_2), ..., (a_k, b_k), (a_{k+1}, b_{k+1}), ... the *scaling process*.

The importance of this process in the present context is a result of the following result:

Lemma 7.9: *Let $(.\geq^{C1}., \geq^{C1})$, $(.\geq^{C2}., \geq^{C2})$, $(.\geq^{C3}., \geq^{C3})$, ..., be a chain in $(\|.\geq.\|, \|\geq\|)$, and suppose that (a_1, b_1), (a_2, b_2), ..., (a_k, b_k), (a_{k+1}, b_{k+1}), ..., is the unique sequence of constants that results from the application of the scaling process when $a_k > 0$ and b_k are chosen arbitrarily. Then, for any indices i and j one has*
7.5*. *$X\|C_i \geq Y\|C_j$ if and only if $a_i V(X\|C_i) + b_i \geq a_j V^{Dj}(Y\|D_j) + b_j$*

Again the proof is presented at the end of this chapter.

Lemma 7.9 and Axiom$_6$ make it possible to represent all the sections in $(\|.\geq.\|, \|\geq\|)$ simultaneously. Axiom$_6$ ensures that $(\|.\geq.\|, \|\geq\|)$ is one big chain. So, if one fixes $a_C > 0$ and b_C for some section $(.\geq^C., \geq^C)$, then the scaling process will produce unique scaling constants a_D and b_D for every other section $(.\geq^D., \geq^D)$. Lemma 7.9 then ensures that the resulting sectional representation $SR^* = \{(P(\bullet\|C), a_C V(\bullet\|C) + b_C): C \in \boldsymbol{C}\}$ is a full representation for $(\|.\geq.\|, \|\geq\|)$.

Thus, the axioms we have set down do perform the job for which they were designed: They capture the notion of conditional evidential expected utility. Since any reasonable decision theory should be expressible in this form, all future work on the foundations of rational choice theory ought to be either attempts to weaken the axioms given here without changing the basic result or attempts to capture special types of conditional evidential expected utility by strengthening the basic axioms presented here.

In this section the central theorems and lemmas of this chapter are proved in the order in which they were presented. It is assumed throughout that the pair $(\|.\geq.\|, \|\geq\|)$ satisfies Axiom$_1$–Axiom$_3$ and that SR $= \{(P(\bullet\|C), V(\bullet\|C)): C \in \boldsymbol{C}\}$ is a *sectional* representation of $(\|.\geq.\|, \|\geq\|)$ whose existence is guaranteed by Lemma 7.2 (which does not itself require proof since it is a consequence of Theorem 4.3). Since SR is a sectional representation of $(\|.\geq.\|, \|\geq\|)$ we can always rely on its being the case that $P(X\|C) \geq P(Y\|C)$ iff $X\|C .\geq. Y\|C$ and that $V(X\|C) \geq V(Y\|C)$ iff $X\|C \geq Y\|C$. We cannot, however, assume that SR is a full representation, in the sense that $P(X\|C) \geq P(Y\|D)$ iff $X\|C .\geq. Y\|D$ and $V(X\|C) \geq V(Y\|D)$ iff $X\|C \geq Y\|D$, since this is what we are trying to prove.

In carrying out the proofs I will make free use of four related facts, all established by Bolker,[14] that pertain to probability/utility pairs that satisfy the Jeffrey/Bolker axioms. All assume a probability/utility representation (P, V) that obeys Jeffrey's Equation on *atomless* algebra Ω.

Fact 1. If $P(X) > 0$ and λ is a real number between 0 and 1, then there is a proposition X^* that entails X and is such that $P(X^*) = \lambda P(X)$ and $V(X) = V(X^*)$.

Fact 2. If $P(X)$ and $P(Y)$ are nonzero and if $V(X) > V(Y)$, then there exist mutually incompatible, nonnull propositions X^* and Y^* that entail X and Y, respectively, and are such that $V(X^*) = V(X) > V(Y^*) = V(Y)$.

Fact 3. If $P(X)$ and $P(Y)$ are nonzero and if $V(X) > V(Z) > V(Y)$, then there exist mutually incompatible, nonnull X^* and Y^* that entail X and Y, respectively, and are such that $V(X^*) = V(X) > V(X^* \vee Y^*) = V(Z) > V(Y^*) = V(Y)$.

Fact 4. V's essential range $I = \{V(X): P(X) > 0\}$ is an interval with nonempty interior on the real line.

Since every $(P(\bullet\|C), V(\bullet\|C))$ pair in SR satisfies Jeffrey's Equation and is defined over an atomless algebra (by Axiom$_3$) these results apply to all of them.

---

[14] See Bolker (1966, lemma 1.17 and lemma 3.5). The two latter facts are fairly obvious consequences of the two former facts.

Here is the first result we need to prove:

Theorem 7.3 (*Existence of Probability Representations for Suppositions*). *If* $\|.\geq.\|$ *and* $\|\geq\|$ *satisfy Axiom$_1$–Axiom$_3$ and SR =* $\{(P(\bullet\|C), V(\bullet\|C)): C \in \mathbf{C}\}$ *is any sectional representation for* $(.\geq^C., \geq^C)$, *then Axiom$_4$ is necessary and sufficient for*

7.1a. SUP$_1$: $P(\bullet\|C)$ *is a countably additive probability on* Ω.
   SUP$_2$: $P(C\|C) = 1$.
   SUP$_3$: $P(X\|C) \geq P(X \& C)$ *when* $C \in \mathbf{C}$.
7.1c. $P(\bullet\|\bullet)$ *ordinally represents* $\|.\geq.\|$.
7.1d. $P(\bullet\|\bullet)$ *is the only function for which 7.1a and 7.1c hold.*

Proof. The necessity of Axiom$_4$ is trivial because each of its conditions is necessary for the existence of a probability representation for $\|.\geq.\|$. So, assume $\|.\geq.\|$ obeys Axiom$_4$. Since each $P(\bullet\|C)$ is a probability for which $P(C\|C) = 1$, and since Axiom$_4$ requires $\|.\geq.\|$ to be regular, there is no question that $P(\bullet\|\bullet)$ is a supposition function. Moreover, since each $.>^C.$ is atomless (by Axiom$_3$) it follows from the Ordinal Uniqueness Theorem of Chapter 4 that $P(\bullet\|C)$ is the *only* representation for $.>^C$, and thus that $P(\bullet\|\bullet)$ must be unique if it is a representation for $\|.\geq.\|$.

To prove 7.1c, let $.\geq^C.$ and $.\geq^D.$ be any two sections of $\|.\geq.\|$. For each integer $n > 0$, Fact 1 guarantees the existence of partitions $\{C(j, n): j = 1, 2, \ldots, n\}$ and $\{D(j, n): j = 1, 2, \ldots, n\}$ such that $P(C(j, n)\|C) = P(D(j, n)\|D) = 1/n$ for all j. Since $P(\bullet\|C)$ and $P(\bullet\|D)$ represent $.>^C.$ and $.>^D.$, this entails that $C(j, n) .=^C. C(k, n)$ and that $D(j, n) .=^D. D(k, n)$ for all j, k, and n. The next step is to show $C(j, n)\|C .=. D(k, n)\|D$ for all cases. If things were otherwise, then (without loss of generality) there would have to be an index n such that

$$C(1,n)\|C .=. C(2,n)\|C .=. \ldots .=. C(n,n)\|C$$
$$.>. D(1,n)\|D .=. D(2,n)\|D .=. \ldots .=. D(n,n)\|D$$

The additivity clause of Axiom$_4$ would then entail that

$$[C(1,n) \vee C(2,n) \vee C(j,n)]\|C$$
$$.>. [D(1,n) \vee D(2,n) \vee D(n,n)]\|D.$$

which contradicts the normality clause of the axiom since the $C(j, n)$'s and $D(j, n)$'s each form a partition. Thus, it must be the case that $C(j, n)\|C .=. D(k, n)\|D$ for all j and k.

Define $C^*(m, n) = [C(1, n) \vee \ldots \vee C(m, n)]$ and $D^*(m, n) = [D(1, n) \vee \ldots \vee D(m, n)]$. The Dominance clause of Axiom$_4$ entails

243

that $C^*(m, n)\|C .=. D^*(m, n)\|D$ for all m and n. Moreover, since the additivity law of probability implies that $P(C^*(m, n)\|C) = P(D^*(j, n)\|D) = m/n$, and since the rational numbers are dense in the reals, both of the following will hold for any $X, Y \in \Omega$:

- $X\|C .>. Y\|C$ iff there are indices m and n with $X\|C .>. C^*(m, n)\|C .>. Y\|C$.
- $X\|D .>. Y\|D$ iff there are indices m and n with $X\|D .>. D^*(m, n)\|D .>. Y\|D$.

If $X\|C .>. Y\|D$ we can use the solvability clause of Axiom$_4$ to find an $X^* \in \Omega$ such that $(X^* \& X)\|C .=. Y\|D$. It then follows that for some indices m and n we have

$$X\|C .>. C^*(m,n)\|C = D^*(m,n)\|D .>. Y\|D$$

and thus that

$$P(X\|C) > P(C^*(m,n)\|C) = m/n = P(D(j,n)\|D) > P(Y\|D)$$

Thus, $X\|C .>. Y\|D$ implies $P(X\|C) > P(Y\|C)$. The converse also holds since there will always be some m and n for which the lower set of relationships holds. This will entail that the upper set of relationships holds as well given that $P(\bullet\|C)$ represents $.\geq^C.$ and that $P(\bullet\|D)$ represents $.\geq^D.$ This completes the proof of Theorem 7.3. ∎

For the next lemma we should remind ourselves that a *test configuration* for $(.\geq^C., \geq^C)$ and $(.\geq^D., \geq^D)$ is a four-tuple $(X_1\|C, X_2\|C, Y_1\|D, Y_2\|D)$ where $X_1 \& X_2$ and $Y_1 \& Y_2$ are both contradictory, and $X_1\|C \approx Y_1\|D > X_2\|C \approx Y_2\|D$. The result we need to prove is

Lemma 7.6. *If $(.\geq^C., \geq^C)$ and $(.\geq^D., \geq^D)$ are linked, then there is at least one test configuration of the form $(X_1\|C, X_2\|C, Y_1\|D, Y_2\|D)$.*

Proof: If $(.\geq^C., \geq^C)$ and $(.\geq^D., \geq^D)$ are linked then there are nonnull pairs $F\|C$, $H\|D$, and $F^*\|C$ such that $F\|C > H\|D > F^*\|C$. Since $\geq^D$ cannot be indifferent among all propositions (Axiom$_3$) there must be an H^* in Ω that is not ranked with H by $\geq^D$. Assume without loss of generality that $H\|D > H^*\|D$. The possibilities then are
(i) $F\|C > H\|D > F^*\|C \geq H^*\|D$
(ii) $H\|C > H\|D > H^*\|D > F^*\|C$
In (i), apply the Solvability clause of Axiom$_5$ twice, once to $\geq^C$ and once to $\geq^D$, to obtain nonnull $Z, Z^* \in \Omega$ with $Z\|C \approx H\|D > F^*\|C$

$\approx Z^*\|D$. In (ii), apply the clause twice to $\geq^C$ to find nonnull Z, $Z^* \in \Omega$ with $Z\|C \approx H\|D > Z^*\|C \approx H^*\|D$. In either event one has $X\|C \approx Y\|D > X^*\|C \approx Y^*\|D$ for some nonnull propositions X, Y, X^*, and Y^*.

Since $V(\bullet\|C)$ represents $\geq^C$ we can apply Fact 2 with $V(X\|C) > V(X^*\|C)$ to find disjoint, nonnull propositions X_1 and X_2 that entail X and X^*, respectively, and are such that $V(X_1\|C) = V(Y\|C) > V(X^*\|C) = V(F^*\|C)$. Doing the same thing with $V(\bullet\|D)$ and $\geq^D$ gives us disjoint, nonnull propositions Y_1 and Y_2 such that $V(Y_1\|D) = V(Y\|D) > V(Y_2\|D) = V(Y^*\|D)$. Since $V(\bullet\|C)$ represents $\geq^C$ and $V(\bullet\|D)$ represents $\geq^D$ this entails $X_1\|C \approx Y_1\|D > X_2\|C \approx Y_2\|D$. $(X_1\|C, X_2\|C, Y_1\|D, Y_2\|D)$ is the test configuration we seek. This completes the proof of Lemma 7.6. ∎

We now turn to the crucial result:

Lemma 7.8. *If* $(.\geq^C., \geq^C)$ *and* $(.\geq^D., \geq^D)$ *are linked, and if* $a_C > 0$ *and* b_C *are chosen arbitrarily and* a_D *and* b_D *are defined by*

7.7. $a_D = a_C[V(X_1\|C) - V(X_2\|C)]/[V(Y_1\|D) - V(Y_2\|D)]$
   $b_D = a_C V(X_1\|C) - a_D V(Y_1\|D) + b_C$

for $(X_1\|C, X_2\|C, Y_1\|D, Y_2\|D)$ *any test configuration associated with* $(.\geq^C., \geq^C)$ *and* $(.\geq^D., \geq^D)$, *then one has*

7.5. $X\|C \geq Y\|D$ *if and only if* $a_C V(X\|C) + b_C \geq a_D V(Y\|D) + b_D$

for all propositions X, $Y \in \Omega$ *such that neither* $X\|C$ *nor* $Y\|D$ *is null. Moreover, the values of* a_D *and* b_D *given by 7.7 are the only ones for which 7.5 holds.*

Proof. To simplify things, let $a_C = 1$ and $b_C = 0$. The proof works the same way with any other choice. What we want to show first is that for any nonnull X and Y it must be the case that

7.5 $X\|C \geq Y\|D$ if and only if $V(X\|C) \geq a_D V(Y\|D) + b_D$

where

7.7 $a_D = [V(X_1\|C) - V(X_2\|C)]/[V(Y_1\|D) - V(Y_2\|D)]$
   $b_D = V(X_1\|C) - a_D V(Y_1\|D)$

We can assume without loss of generality that $X\|C \geq Y\|D$, so the goal will be to establish

7.5a $V(X\|C) \geq a_D V(Y\|D) + b_D$

The proof will be broken into cases depending on where $X\|C$ and $Y\|D$ fall in relation to elements of the test configuration.

In working through the cases one must keep in mind that the test configuration is such that $X_1 \| C \approx Y_1 \| D > X_2 \| C \approx Y_2 \| D$, and that the constants a_D and b_D have been specifically chosen to ensure that $V(X_1 \| C) = a_D V(Y_1 \| D) + b_D$ and $V(X_2 \| C) = a_D V(Y_2 \| D) + b_D$.

Case 1. (1a) $X \| C \geq X_1 \| C \approx Y_1 \| D \geq Y \| D$
(1b) $X \| C \geq X_2 \| C \approx Y_2 \| D \geq Y \| D$

In subcase 1a one has $V(X \| C) \geq V(X_1 \| C) = a_D V(Y_1 \| D) + b_D \geq a_D V(Y \| D) + b_D$. Subcase 1b is identical with X_1 and Y_1 replaced by X_2 and Y_2.

Case 2. $X_1 \| C \approx Y_1 \| D$. $X \| C \geq Y \| D$. $X_2 \| C \approx Y_2 \| D$.

Here we appeal to Fact 3 twice to find nonnull X_1^*, Y_1^*, X_2^*, $Y_2^* \in \Omega$ that entail X_1, Y_1, X_2, and Y_2, respectively, and are such that

$$X_1^* \| C \approx Y_1^* \| D \approx X_1 \| C$$
$$(X_1^* \vee X_2^*) \| C \approx X \| C$$
$$(Y_1^* \vee Y_2^*) \| D \approx Y \| D$$
$$X_2^* \| C \approx Y_2^* \| D \approx X_2 \| C.$$

Assume, without loss of generality, that $(X_1^* \vee X_2^*) \| C \geq (Y_1^* \vee Y_2^*) \| D$ and set

$$\lambda = P(Y_1^* \vee Y_2^* \| D) / P(X_1^* \vee X_2^* \| C)$$

(which is sure to be well defined because X_1^* and X_2^* are nonnull in the C-section). Since $1 \geq \lambda > 0$ we can apply Bolker's Fact 1 to find nonnull X_1^{**}, $X_2^{**} \in \Omega$ that entail X_1^* and X_2^*, respectively, and are such that

$$V(X_1^{**} \| C) = V(X_1^* \| C) \text{ and } P(X_1^{**} \| C) = \lambda P(X_1^* \| C)$$
$$V(X_2^{**} \| C) = V(X_2^* \| C) \text{ and } P(X_2^{**} \| C) = \lambda P(X_2^* \| C)$$

This entails that $V(X \| C) = V(X_1^{**} \vee X_2^{**} \| C)$ since

$$V(X \| C) = V(X_1^* \vee X_2^* \| C)$$
$$= \frac{P(X_1^* \| C)}{P(X_1^* \vee X_2^* \| C)} V(X_1^* \| C) + \frac{P(X_2^* \| C)}{P(X_1^* \vee X_2^* \| C)} V(X_2^* \| C)$$
$$= \frac{\lambda P(X_1^{**} \| C)}{\lambda P(X_1^{**} \vee X_2^{**} \| C)} V(X_1^{**} \| C)$$
$$+ \frac{\lambda P(X_2^{**} \| C)}{\lambda P(X_1^{**} \vee X_2^{**} \| C)} V(X_2^{**} \| C)$$

246

The first identity follows from the indifference $(X_1^* \vee X_2^*)\|C \approx X\|C$ and the fact that $V(\bullet\|C)$ represents $\geq^C$. The third is a consequence of the identities $V(X_1^{**}\|C) = V(X_1^*\|C)$ and $P(X_1^{**}\|C) = \lambda P(X_1^*\|C)$, and the fact that X_1^{**} and X_2^{**} are mutually incompatible. The other two identities hold because $V(\bullet\|C)$ obeys Jeffrey's Equation.

Since $V(\bullet\|D)$ represents $\geq^D$ it is also true that $V(Y\|D) = V(Y_1^* \vee Y_2^*\|D)$. So, we can establish the desired inequality 7.5a by proving

7.5b $V(X_1^{**} \vee X_2^{**}\|C) \geq a_D V(Y_1^* \vee Y_2^*\|D) + b_D$.

Start by using the identities $V(X_1^{**}\|C) = V(X_1\|C)$, $V(X_2^{**}\|C) = V(X_2\|C)$, $V(Y_1^*\|D) = V(Y_1\|D)$, and $V(Y_2^*\|D) = V(Y_2\|D)$ to rewrite 7.7 as

7.7* $a_D = [V(X_1^{**}\|C) - V(X_2^{**}\|C)]/[V(Y_1^*\|D) - V(Y_2^*\|D)]$
$b_D = V(X_1^{**}\|C) - a_D V(Y_1^*\|D)$

This (and a little algebra) allows us to express 7.5b as

$$\frac{V(X_1^{**} \vee X_2^{**}\|C) - V(X_1^{**}\|C)}{V(X_1^{**}\|C) - V(X_2^{**}\|C)} \geq \frac{V(Y_1^* \vee Y_2^*\|C) - V(Y_1^*\|C)}{V(Y_1^*\|C) - V(Y_2^*\|C)}$$

Since $V(\bullet\|C)$ and $V(\bullet\|D)$ obey Jeffrey's Equation in conjunction with $P(\bullet\|C)$ and $P(\bullet\|D)$, respectively, we can rewrite this inequality in terms of probabilities as

$$\frac{P(X_2^{**}\|C)}{P(X_1^{**} \vee X_2^{**}\|C)} \leq \frac{P(Y_2^*\|D)}{P(Y_1^* \vee Y_2^*\|D)}$$

(Note the change in the direction of the inequality.)

Since X_1^{**} and X_2^{**} were chosen to make it true that $P(X_1^{**}\|C) = \lambda P(X_1^*\|C)$ and $P(X_2^{**}\|C) = \lambda P(X_2^*\|C)$ where $\lambda = P(Y_1^* \vee Y_2^*\|D)/P(X_1^* \vee X_2^*\|C)$ it follows that $P(X_1^{**} \vee X_2^{**}\|C) = \lambda P(X_1^* \vee X_2^*\|C) = P(Y_1^* \vee Y_2^*\|D)$. This allows us to simplify the inequality we need to establish still further to

$$P(X_2^{**}\|C) \leq P(Y_2^*\|D)$$

and thus to

$$X_2^{**}\|C \leq. Y_2^*\|D$$

since $P(\bullet\|\bullet)$ ordinally represents $\|.>\|$.

To see why this last relationship has to hold notice that we have a situation in which

- X_1^{**} and X_2^{**} are nonnull, mutually incompatible propositions
- Y_1^* and Y_2^* are nonnull, mutually incompatible propositions
- $X_1^{**}\|C \approx Y_1^*\|D > (X_1^{**} \vee X_2^{**})\|C \geq (Y_1^* \vee Y_2^*)\|C > X_2^{**}\|C \approx Y_2^*\|D$

This shows that $(X_1^{**}\|C, X_2^{**}\|C, Y_1^*\|D, Y_2^*\|D)$ is a test configuration. Moreover, since we know that $(X_1^{**} \vee X_2^{**})\|C.=. (Y_1^* \vee Y_2^*)\|D$ it follows from the last clause of the Generalized Coherence principle, Axiom$_7$, that $X_1^{**}\|C .\geq. Y_1^*\|D$, and this implies that $X_2^{**}\|C .\leq. Y_2^*\|D$. 7.5a is thus established in Case 2.

Case 3. $X\|C \geq Y\|D > X_1\|C \approx Y_1\|D > X_2\|C \approx Y_2\|D$.

The proof here is similar to that of Case 2, so I will merely sketch the main ideas. Note first that when $X\|C > Y\|D$ we can use the Solvability clause of Axiom$_5$ to find a nonnull X^* such that $X^*\|C \approx Y\|D$. Thus, we can assume without loss of generality that we are dealing with a case in which $X\|C \approx Y\|D$ because if we show that $V(X\|C) = a_D V(Y\|D) + b_D$ in this case then $V(Z\|C) > a_D V(Y\|D) + b_D$ will follow whenever $V(Z\|C) > V(X\|C)$. Our goal, then, is to establish that $V(X\|C) = a_D V(Y\|D) + b_D$ given that $X\|C \approx Y\|D > X_1\|C \approx Y_1\|D > X_2\|C \approx Y_2\|D$.

Using Fact 3 in essentially the same way as in Case 2 we can find propositions X^*, Y^*, X_2^*, $Y_2^* \in \Omega$ that entail X, Y, X_2, and Y_2, respectively, and are such that

- X^* and X_2^* are mutually incompatible
- Y^* and Y_2^* are mutually incompatible
- $X^*\|C \approx X\|C \geq Y^*\|D \approx Y\|D > (X^* \vee X_2^*)\|C \approx X_1\|C \approx (Y^* \vee Y_2^*)\|D \approx Y_1\|D > X_2^*\|C \approx Y_2^*\|D \approx X_2\|C \approx Y_2\|D$.

Again, we can assume, without loss of generality, that $(X^* \vee X_2^*)\|C .\geq. (Y^* \vee Y_2^*)\|D$, set

$$\lambda = P(Y^* \vee Y_2^*\|D)/P(X^* \vee X_2^*\|C)$$

and then use Fact 1 to find nonnull X^{**}, $X_2^{**} \in \Omega$ that entail X^* and X_2^*, respectively, and are such that

$$V(X^{**}\|C) = V(X^*\|C) \text{ and } P(X^{**}\|C) = \lambda P(X^*\|C)$$
$$V(X_2^{**}\|C) = V(X_2^*\|C) \text{ and } P(X_2^{**}\|C) = \lambda P(X_2^*\|C)$$

Reasoning analogous to that used in Case 2 yields that $V(X_1\|C) = V(X^{**} \vee X_2^{**}\|C)$, and it follows that $(X^{**}\|C, X_2^{**}\|C, Y^*\|D, Y_2^*\|D)$ is a test configuration in which

$$X^{**}\|C \approx Y^*\|D > (X^{**} \vee X_2^*)\|C \approx (Y^* \vee Y_2^*)\|D > X_2^{**}\|C \approx Y_2^*\|D.$$

Now, since $V(X^{**}\|C) = V(X\|C)$ and $V(Y^*\|D) = V(Y\|D)$, the relevant version of 7.5a for this case can be established by proving that $V(X^{**}\|C) = a_D V(Y^*\|D) + b_D$. Since $V(X^{**} \vee X_2^*\|C) = V(X_1\|C)$, $V(X_2^{**}\|C) = V(X_2\|C)$, $V(Y^* \vee Y_2^*\|D) = V(Y_1\|D)$, and $V(Y_2^*\|C)$

248

= V($Y_2\|C$), we can use 7.7 and Jeffrey's Equation to rewrite this inequality as

$$\frac{V(X^{**}\|C) - V(X^{**} \vee X_2^{**}\|C)}{V(X^{**} \vee X_2^{**}\|C) - V(X_2^{**}\|C)} = \frac{V(Y^*\|D) - V(Y^* \vee Y_2^*\|D)}{V(Y^* \vee Y_2^*\|D) - V(Y_2^*\|C)}$$

And, in the presence of Jeffrey's Equation, this is equivalent to

$$P(X^{**}\|C)/P(X_2^{**}\|C) = P(Y^*\|C)/P(Y_2^*\|C)$$

Given that $P(X^{**}\|C) + P(X_2^{**}\|C) = \lambda P(X^* \vee X_2^*\|C) = P(Y^* \vee Y_2^*\|D)$, this holds if and only if $P(X_2^{**}\|C) = P(Y_2^*\|C)$ or, equivalently, $Y_2^*\|C \doteq X_2^{**}\|C$ (since $P(\bullet\|\bullet)$ represents $\|..>\|$). This follows from Axiom$_7$ because $X^{**} \vee X_2^*\|C \doteq Y^* \vee Y_2^*\|D$ and $(X^{**} \vee X_2^*)\|C \approx (Y^* \vee Y_2^*)\|D$. This completes the proof of 7.5a for Case 3. ∎

Case 4. $X_1\|C \approx Y_1\|D > X_2\|C \approx Y_2\|D > X\|C \geq Y\|D$.

This is almost identical to Case 3, and I leave it to the reader.

Since Cases 1–4 exhaust the possibilities, we have shown that 7.5a holds when a_D and b_D are given by 7.7. Showing that no other values do the job is a matter of noting that if both $V(X_1\|C) = aV(Y_1\|D) + b$ and $V(X_2\|C) = aV(Y_2\|D) + b$ are going to hold, then $b = V(X_1\|C) - aV(Y_1\|D) = V(X_2\|C) - aV(Y_2\|D)$, and from this it follows directly that $a = [V(X_1\|C) - V(X_2\|C)]/[V(Y_1\|D) - V(Y_2\|D)]$. This completes the proof of Lemma 7.8.

We turn now to the final lemma we shall need to prove. It assumes a chain of sections $(.\geq^{C1}., \geq^{C1}), (.\geq^{C2}., \geq^{C2}), (.\geq^{C3}., \geq^{C3}), \ldots$. To simplify the presentation I am going to write these as $(.\geq_1., \geq_1), (.\geq_2., \geq_2), (.\geq_3., \geq_3), \ldots$, and denote their utilities from SR by $V_j(\bullet) = V_j(\bullet \|C_j)$ for $j = 1, 2, 3, \ldots$.

Lemma 7.9. *Let $(.\geq_1., \geq_1), (.\geq_2., \geq_2), (.\geq_3., \geq_3), \ldots$ be a chain of sections in $(\|.\geq.\|, \|\geq\|)$, and suppose that $(a_1, b_1), (a_2, b_2), \ldots, (a_k, b_k), (a_{k+1}, b_{k+1}), \ldots$, is the unique sequence of constants that results from the application of the scaling process when $a_k > 0$ and b_k are chosen arbitrarily, so that*

$$a_{j+1} = [V_j(X_j) - V_j(X_j^*)]/[V_{j+1}(Y_j) - V_{j+1}(Y_j^*)]$$
$$b_{j+1} = V_j(X_j) - a_{j+1}V_{j+1}(Y_j^*)$$

where $(X_j\|C_j, X_j^*\|C_j, Y_j\|C_{j+1}, Y_j^*\|C_{j+1})$ is a test configuration for each j. Then, for any indices j and k one has

249

7.5* $X\|C_j \geq Y\|C_k$ if and only if $a_jV_j(X) + b_j \geq a_kV_k(Y) + b_k$

Proof. Assume that $k > j$. The first step is to "thin out" the chain between $(.\geq_{j.}, \geq_j)$ and $(.\geq_{k.}, \geq_k)$ so that each section is linked only to its immediate predecessor and successor. There is a simple algorithm for doing this: Go to the least m such that $(.\geq_{m+1.}, \geq_{m+1})$ and $(.\geq_{m-1.}, \geq_{m-1})$ are linked; throw away the section $(.\geq_{m.}, \geq_m)$; repeat the process until it is no longer possible to do so. This procedure always leaves us with a chain that has $(.\geq_{j.}, \geq_j)$ as its first link, $(.\geq_{k.}, \geq_k)$ as its last link, and in which one section is never linked to the section that follows its immediate successor. To keep the proof simple we will simply assume that we had a chain of this sort to begin with.

Fact 4 implies that the essential range $I_m = \{a_mV_m(X) + b_m: P(X\|C_m) > 0\}$ of each utility function $a_mV_m(X) + b_m$ ($m = j, j + 1, \ldots, k - 1$) is an interval on the real line with nonempty interior. The previous lemma entails that the sections $(.\geq_{m.}, \geq_m)$ and $(.\geq_{n.}, \geq_n)$ are linked if and only if I_m and I_n overlap in an interval with nonempty interior. Thus, since the chain we are dealing with is thin, the intersection of I_m and I_{m+2} must be empty for all $m = j, j + 1, \ldots, k - 1$. The only way for this to happen when each $I_m \cap I_{m+1}$ is nonempty is for there to be a real number z_m in each I_m such that either

(I) $x_m \geq z_m \geq y_m$ for all $x_m \in I_{m-1}, y_m \in I_{m+1}$

or

(II) $x_m \leq z_m \leq y_m$ for all $x_m \in I_{m-1}, y_m \in I_{m+1}$

The intervals, in other words, must be overlapping and *descending*, as in case (I), or overlapping and *ascending*, as in case (II). Without loss of generality assume that (I) is the relevant possibility. Given that each z_m is in the essential range of $a_mV_m(\bullet) + b_m$ this means that there must be propositions $Z_j, Z_{j+1}, \ldots, Z_k$ such that

(#) $X\|C_{m-1} \geq Z_m\|C_m \geq Y\|C_{m+1}$ for all $X, Y \in \Omega$ and $m = j, \ldots, k - 1$.

And, since we already know from the previous lemma that 7.5* holds for all when $k = j + 1$ we also have

(##) $a_{m-1}V_{m-1}(X) + b_{m-1} \geq a_mV_m(Z_m) + b_{m+1}$
$\geq a_{m+1}V_{m+1}(Y) + b_{m+1}$ for all $X, Y \in \Omega$ and $m = j, \ldots, k - 1$.

Now, to complete the proof merely notice that we will either have $k = j + 1$ in which case the chain has only one link and 7.5* follows from the lemma. Or, on the other hand, if $k > j + 1$ then (#) and (##) imply

$$X\|C_j \geq Z_{j+1}\|C_{j+1} \geq Y\|C_k \text{ for all } X, Y \in \Omega$$

and

$$a_j V_j(X) + b_j \geq a_{j+1} V_{j+1}(Z_{j+1}) + b_{j+1} \geq a_k V_k(Y) + b_k$$
$$\text{for all } X, Y \in \Omega$$

7.5* follows directly. This completes the proof of Lemma 7.9. ∎

8

Where Things Stand

There are two general lessons that I would like readers to draw from the foregoing chapters. First, it should now be clear that all expected utility theorists can agree about the broad foundational assumptions that underlie their common doctrine. In particular, the representation theorem of Chapter 7 should remove any lingering doubts there might have been about the theoretical underpinnings of causal decision theory. Since the constraints on conditional preferences and beliefs needed to establish the existence of conditional utility representations in Theorem 7.4 are common to both the causal and evidential theories, there is really no difference between them as far as their core accounts of *valuing* are concerned. If asked the question, "Given a fixed system of beliefs and basic desires, what is the right way to assign utilities to risky or uncertain prospects?" proponents of both positions should answer in exactly the same way, by saying that a prudentially rational agent will use the specified beliefs and desires to compute a news value for each prospect, and that her preferences should go by increasing auspiciousness. In the end, then, Richard Jeffrey was right about the nature of value and Ethan Bolker was correct about the way in which theories of value should be characterized formally. These are points to which all parties should be able to agree.

There remains, of course, an important difference between the causal and evidential approaches to decision theory. Even though they agree about the way in which prospects should be valued once an epistemic perspective is in place, the two theories differ about the correct epistemic perspective from which an *agent* should evaluate his or her potential actions. According to the causal theory, which I have endorsed, a rational decision maker is no mere passive conduit of behavior who assesses her deliberate acts relative to the same standard that she uses to evaluate events she cannot control. She is, rather, a genuine *agent* who aims to *produce* good, not merely to produce evidence for it, and thus who evaluates her options solely on the basis

252

of what they are likely to causally promote. The rational agent, there-fore, will adopt the epistemic perspective appropriate for taking the specifically causal properties of her actions into account. As has been explained, this involves first supposing, in the *subjunctive* sense, that she performs each act and then determining its value as news when it is viewed from that perspective. The evidential decision theorist's error lies in thinking that differences in news values computed relative to a person's *ordinary* unconditional probabilities can provide her with reasons for acting. The fact is that preferences for news provide reasons for action only when they are formulated on the basis of an epistemic perspective that makes the causal properties of acts manifest.

Thus the first major lesson that readers should take away from this work is that the evidential decision theorists have been right all along about the nature of rational desire, but they have mistakenly thought that all desires provide reasons for action. The fact that A would be better news than B does not give an agent a reason to choose A over B unless what is meant is that A's news value *on the subjunctive supposition that it is performed* is greater than B's news value *on the subjunctive supposition that it is performed*. The moral, then, is that Jeffrey's theory is not really a logic of *decision* but a logic of *rational desire*. It becomes a logic of decision only when it is applied in the right frame of mind, the frame of mind appropriate to someone who is concerned about what she might cause.

A second broad lesson is that the pragmatist ideology that has driven research in decision theory since its inception must be aban-doned. The results obtained here – the uniqueness theorem for evi-dential decision theory in Chapter 4 and the general representation result of Chapter 7 – were possible only because we were willing to introduce *independent* laws of rationality to govern beliefs and de-sires. We have seen repeatedly that the orthodox approach, in which canons of rational belief are derived from canons of rational desire, either requires dubious technical devices, like "constant" actions or "mitigators," or leaves us (as in the case of the Jeffrey/Bolker theory) with an unacceptable level of indeterminacy in our expected utility representations. This is no accident; the laws of rational desire *should* be insufficient to fix the laws of rational belief because beliefs answer to a *norm of accuracy* that has no direct analogue for desires. As Elizabeth Anscombe noted long ago,[1] the standard of success for a belief is that it should "fit the world." A rational agent who believes that P and who receives evidence that tells against P thereby acquires

---

[1] Anscombe (1957).

a (defeasible) reason to lower her degree of confidence in P. Desires have an altogether different "direction of fit." Unlike beliefs, they are not supposed to fit the world but, whenever possible, to be fit *by* it. An agent who desires P and receives evidence that tells in favor of $\neg P$ thereby acquires a (defeasible) reason not to change her desire, but to change the world so as to bring about P (if such a thing is within her power). Given these two rather different "job descriptions" it is not surprising that the laws of rational belief do not simply fall out of the laws of rational desire.

The second major lesson, then, is that decision theory must throw off the pragmatist/behaviorist straitjacket that has hindered its progress for the past seventy years. Doing so will allow decision theorists the freedom to formulate their theories in terms of constraints on rational belief as well as rational desire, and (as the results obtained here attest) this is sure to lead to a deeper understanding of the decision making process.

8.2 THREE AREAS FOR FUTURE RESEARCH

There is a great deal that has been left undone. Three issues seem particularly pressing. First, the remarks at the end of the previous section suggested that rational beliefs answer to a "norm of accuracy" that does not apply to rational desires. This raises a host of questions. What does it mean to say that beliefs must "fit the world"? How exactly is accuracy or "fit" to be understood in this context? Can it be measured? Why think that the axiomatic requirements that we have imposed on beliefs should make them more accurate; that is, does probabilistic coherence enhance the degree of "fit" between beliefs and the world? These are all important questions that deserve serious answers, but since they are ultimately matters of epistemology rather than decision theory proper I have chosen not to address them here. This is one major piece of unfinished business.[2]

A second is that all existing representation theorems for expected utility theory, including those established here, make use of strong "structure" axioms that are not strictly necessary for the existence of the desired representation. In Theorem 7.4, for example, we needed to assume (as a part of Axiom$_3$) that every section of the agent's likelihood ranking splits each nonnull proposition into an infinity of mutually incompatible nonnull parts. At a number of points I have endorsed the prevailing view, which interprets structure axioms as

---

[2] See my (1998) for answers to these questions.

254

requirements of coherent extendibility. It must be admitted, however, that this is something of a dodge. As noted in Chapter 3, the ultimate aim of foundational work in decision theory should be to discover a set of intuitively compelling axiomatic constraints on preferences and comparative beliefs, each of which is individually necessary for representability and that together suffice to guarantee the existence of an expected utility representation. Thus, to obtain a fully satisfactory representation theorem for abstract conditional decision theory and its special cases of causal and evidential decision theory, we must seek to replace the structure axioms that appear in Theorem 7.4 with "axioms of pure rationality" that need not be understood as requirements of coherent extendibility. This, as noted in Chapter 3, would be tantamount to proving a "completeness theorem" for expected utility theory. The problem is to find the right axioms to do the job.

I have made a few preliminary forays into this territory, but without much success. Given what has been accomplished in the preceding chapters, the problem reduces to one of finding a set of axioms that are both (a) necessary and sufficient for representability *within evidential decision theory*, and (b) plausible as requirements of rationality. Zolton Domotor has presented a set of axioms that do part of the first job,[3] but that, as far as I am able to determine, fail rather badly at the second. He proves a representation theorem that establishes necessary and sufficient conditions for the existence of a *non-Archimedean* representation for preferences defined on finite algebras of propositions. (The non-Archimedean nature of the representation is a plus, but the restriction to finite algebras is not.) The main axiom used in the proof encodes a very complicated infinite family of conditions, each expressed in terms of characteristic functions of *Cartesian products* of propositions (viewed as sets of possible worlds). Its simplest instance would be difficult to write down on this page. The point of the axiom is to ensure that, for any sequences of propositions $(A_1, A_2, \ldots, A_n)$, $(B_1, B_2, \ldots, B_n)$, $(C_1, C_2, \ldots, C_n)$, and $(D_1, D_2, \ldots, D_n)$, if one has

- $0 = \Sigma_{j \leq n} P(A_j)P(B_j)P(C_j)P(D_j)[u(B_j) - u(A_j)][u(D_j) - u(C_j)]$
- $u(B_j) \geq u(A_j)$ for all $j \leq n$ and $u(D_j) \geq u(C_j)$ for all $j < n$.
- $0 = P(A_n)P(B_n)P(C_n)P(D_n)\{[u(B_n) - u(A_n)][u(D_n) - u(C_n)] + [u(C_n) - u(A_n)][u(B_n) - u(D_n)] + [u(D_n) - u(A_n)][u(C_n) - u(B_n)]\}$

then one also has $u(C_n) \geq u(D_n)$. Interested readers are invited to try to figure out some reason why this should be a requirement of ration-

[3] Domotor (1978).

ality. I myself side with Domotor, who frankly admits that his axiom "does not have much intuitive appeal, no matter how long one contemplates it."[4] While he correctly points out that many of the consequences of his axiom are reasonable, this cannot fully compensate for its almost incomprehensible complexity. After all, the reasonable consequences he adduces are not themselves sufficient to yield the desired result. Moreover, even if they were, they also impose constraints that involve algebraic relations among characteristic functions of Cartesian products of propositions, and it is hard to see how any such condition is going to issue in plausible norms of rational desire.

My hunch is that we will be able to make headway on the problem of finding necessary and sufficient conditions for representability within evidential decision theory by employing axioms that constrain both preferences and comparative probability judgments.[5] What looks absurd as a restriction on rational preference might turn out to be perfectly sensible when viewed as a constraint on belief. I suspect that the desired result can be obtained by modifying Theorem 4.3, my reformulation of Bolker's theorem, so that (i) all its structure axioms are deleted, (ii) its nonstructure axioms pertaining exclusively to preferences (EDT_1, EDT_2, EDT_4–EDT_9) are left intact, (iii) the Coherence principle is strengthened, and (iv) all the constraints on comparative probabilities CP_1–$*CP_8$, except the continuity axiom CP_7, are replaced by the following condition due to Dana Scott:

Scott's Axiom.[6] *If a pair of sequences of (not necessarily distinct) propositions $(X_1, \ldots, X_n)$ and $(Y_1, \ldots, Y_m)$ are isovalent in the sense that the number of truths in the first is sure to be identical to the number of truths in the second as a matter of logic, then it should never be the case that $X_i .\geq. Y_i$ for all $i = 1, 2, 3, \ldots, n$ with $X_j .>. Y_j$ for some j.*

Scott's Axiom (plus a nontriviality requirement) turns out to be both necessary and sufficient for the existence of a probability representation for $(.>., .\geq.)$. Proving that a corresponding cohering utility exists will, I conjecture, be a matter of carrying out task (iii) by finding a suitably general version of Coherence. While I have some vague guesses about what such a generalization might look like, I have nothing hard to go on at the moment. The problem of finding

[4] Domotor (1978, p. 176).
[5] This is a part of what Jeffrey suggested in the passage quoted at the end of sec. 4.3.
[6] This condition was first formulated, in a less perspicacious way, by Kraft, Pratt, and Seidenberg in their (1959). Scott's reformulation may be found in his (1964).

256

intuitively plausible necessary and sufficient conditions for representability thus remains unsolved. It is very definitely a problem that merits further study.

A final major piece of unfinished business concerns the appropriate formal model to use in representing subjunctive suppositions. Chapter 6 contained a generalization of Lewis's notion of imaging that defined the image of a probability function P on condition C as $P(X\backslash C) = \Sigma_W P(W)\rho^C(X, W)$ where the function ρ^C is a probability that specifies the proportion of W's probability that gets shifted to X in the move from P to $P(\bullet\backslash C)$. This function was required to satisfy two constraints that related its values to the agent's judgments about similarity among possible worlds. First there was a kind of dominance principle that required $\rho^C(X, W)$ to be larger than $\rho^C(Y, W)$ if every C-world in which X holds is more like W than any world in which Y holds. The second requirement was that $\rho^C(X, W)$ has to be 1 if it is impossible to partition X into nonempty, mutually incompatible propositions X_1 and X_2 such that every X_1-world is more like W than any X_2-world. I consider these the *minimum* requirements that $\rho^C(X, W)$ should satisfy, and I suspect that further reflection on the relationship between image probabilities and comparative judgments of similarity among possible worlds will lead to more substantive and illuminating constraints.

Moreover, in my discussion of imaging I said next to nothing about the nature of the similarity relation itself, and I was careful to make sure that the conditions imposed on ρ would hold good for any reasonable similarity relation so as not the prejudice the issue. As David Lewis noted,[7] however, the similarity among worlds needs to be understood in a special, "nonbacktracking" manner if subjunctive supposition is going to be able to capture the sorts of beliefs about causal relations that are relevant to decision making. Thus, not every proportion function ρ will give rise to a form of belief revision that is implicated in the process of choosing actions. What one needs to investigate is how a fuller understanding of the notion of "nonbacktracking" similarity will further constrain the admissible ρ functions.

However these problems are eventually solved, one thing we can be certain about is that any reasonable decision theory will be causal; it will explain and justify actions by making explicit reference to the actor's beliefs concerning what she has the power to bring about by an exercise of her will. To think otherwise is to miss the fact that the whole point of engaging in deliberate decision making is to change the world for the better.

---

[7] Lewis (1979).

References

Adams, Ernest. 1965. The Logic of Conditionals. *Inquiry* 8: 166–97.

Anderson, Elizabeth. 1993. *Value in Ethics and Economics*. Cambridge: Harvard University Press.

Anscombe, G.E.M. 1957. *Intention*. Oxford: Basil Blackwell.

Armendt, Brad. 1986. A Foundation for Causal Decision Theory. *Topoi* 5: 3–19.

Arnauld, Antoine and Pierre Nicole. [1662]/1996. *Logic or the Art of Thinking, 5th ed.* Translated and edited by Jill Vance Buroker. Cambridge: Cambridge University Press.

Arrow, Kenneth. 1966. Exposition of the Theory of Choice under Uncertainty. *Synthese* 16: 253–69.

Bennett, Jonathan. 1988. Farewell to the Phlogistron Theory of Conditionals. *Mind* 97: 509–27.

Bolker, Ethan. 1966. Functions Resembling Quotients of Measures. *Transactions of the American Mathematical Society* 124: 293–312.

Bradley, Richard. 1998. A Representation Theorem for a Decision Theory with Conditionals. *Synthese* 116, no. 2.

Bratman, Michael. 1987. *Intentions, Plans and Practical Reason*. Cambridge: Harvard University Press.

Broome, John. 1991. *Weighing Goods*. Oxford: Blackwell.

Carnap, Rudolf. 1936. Testability and Meaning. *Philosophy of Science* 3: 419–71.

Cartwright, Nancy. 1979. Causal Laws and Effective Strategies. *Noûs* 13: 419–37.

Christensen, David. Modeling Quantitative Confirmation: A New Take on Old Evidence. Unpublished manuscript.

Diaconis, Persi and Sandy Zabell. 1982. Updating Subjective Probability. *Journal of the American Statistical Association* 77: 822–30.

de Finetti, Bruno. 1964. Foresight: Its Logical Laws, Its Subjective Sources. In *Studies in Subjective Probability*, edited by H. Kyburg and H. Smokler, pp. 93–158. New York: John Wiley.

Doring, Frank. A Decision Calculus Based on Conditional Probabilities. Unpublished manuscript.

Duff, Anthony. 1986. Pascal's Wager and Infinite Utilities. *Analysis* 46: 107–09.

Diamond, Peter. 1967. Cardinal Welfare, Individualistic Ethics, and Interpersonal Comparisons of Utility: Comment. *Journal of Political Economy* 75: 765–6

Earman, John. 1992. *Bayes or Bust.* Cambridge, MA: MIT Press.

Eells, Ellery. 1982. *Rational Decision and Causality.* Cambridge: Cambridge University Press.

Eells, Ellery. 1985a. Problems of Old Evidence. *Pacific Philosophical Quarterly* 66: 283–302.

Eells, Ellery. 1985b. Levi's Wrong Box. *Journal of Philosophy* 82: 91–104.

Eells, Ellery. 1991. *Probabilistic Causality.* New York: Cambridge University Press.

Fine, Terrence. 1973. *Theories of Probability.* New York: Academic Press.

Fishburn, Peter. 1970. *Utility Theory for Decision Making.* New York: Wiley.

Fishburn, Peter. 1973. A Mixture-Set Axiomatization of Conditional Subjective Expected Utility. *Econometrica* 41: 1–25.

Fishburn, Peter. 1974. On the Foundations of Decision Making under Uncertainty. In *Essays on Economic Behavior under Uncertainty,* edited by M. Balch, D. McFadden, and S. Wu, pp. 1–25. Amsterdam: North Holland.

Friedman, K. and A. Shimony. 1971. Jaynes Maximum Entropy Prescription and Probability Theory. *Journal of Statistical Physics* 3: 381–84.

Gaifman, Haim. 1988. A Theory of Higher Order Probability. In *Causality, Chance and Choice,* edited by B. Skyrms and W. Harper, pp. 191–219. Dordrecht: Kluwer.

Garber, Daniel. 1983. Old Evidence and Logical Omniscience in Bayesian Confirmation Theory. In *Testing Scientific Theories,* edited by J. Earman, *Midwest Studies in the Philosophy of Science,* vol. X, pp. 99–131. Minneapolis: University of Minnesota Press.

Gardenfors, Peter. 1982. Imaging and Conditionalization. *Journal of Philosophy* 79: 747–60.

Gardenfors, Peter. 1988. *Knowledge in Flux.* Cambridge: MIT Press.

Gibbard, Allan. 1980. Two Recent Theories of Conditionals. In *IFS: Conditionals, Belief, Decision, Chance and Time,* edited by W. Harper, R. Stalnaker, and G. Pierce, pp. 211–47. Dordrecht: Reidel.

Gibbard, Allan. 1984. Decision Matrices and Instrumental Expected Utility. Paper for the *Second International Congress on the Foundations of Utility,* Venice, Italy, June 1984.

Gibbard, Allan and William Harper. 1978. Counterfactuals and Two Kinds of Expected Utility. In *Foundations and Applications of Decision Theory,* edited by C. Hooker, J. Leach, and E. McClennen, pp. 125–62. Dordrecht: Reidel. (Reprinted with revisions in P. Gardenfors and N. Sahlin, eds. *Decision, Probability and Utility,* pp. 340–76. Cambridge: Cambridge University Press, 1988.)

Glymour, Clark. 1980. *Theory and Evidence.* Princeton: Princeton University Press.

Hajek, Alan. Waging War on Pascal's Wager. Unpublished manuscript.

Hajek, Alan and Ned Hall. 1994. The Hypothesis of the Conditional Construal of Conditional Probability. In *Probability and Conditionals,* edited by E. Eells and B. Skyrms, pp. 75–112. New York: Cambridge University Press.

Hammond, Peter. 1994. Elementary non-Archimedean Representations for of Probability for Decision Theory and Games. In *Patrick Suppes: Scientific Philosopher*, vol. 1, edited by P. Humphreys, pp. 25–62. Dordrecht: Kluwer.

Harman, Gilbert. 1973. *Thought*. Princeton: Princeton University Press.

Hargreaves Heap, Shaun, Martin Hollis, Bruce Lyons, Robert Sugden, and Albert Weale. 1992. *The Theory of Choice: A Critical Guide*. Oxford: Basil Blackwell.

Harper, William. 1976. Rational Belief Change, Popper Functions and Counterfactuals. In *Foundations of Probability Theory, Statistical Inference, and Statistical Theories of Science*, vol. I, edited by W. Harper and C. Hooker, pp. 73–115. Dordrecht: Reidel.

Harper, William and Brian Skyrms, eds. 1988. *Causation in Decision, Belief Change, and Statistics, II*. New York: Kluwer Academic Press.

Hitchcock, Christopher. 1993. A Generalized Probabilistic Theory of Causal Relevance. *Synthese* 97(3): 335–64.

Hitchcock, Christopher. 1996. Causal Decision Theory and Decision-Theoretic Causation. *Noûs*: 508–26.

Hobson, A. 1971. *Concepts in Statistical Mechanics*. New York: Gordon and Breach.

Hofstadter, Douglas. 1983. Metamagical Themas. *Scientific American*, June.

Horgan, Terrence. 1981. Counterfactuals and Newcomb's Problem. *Journal of Philosophy* 68: 331–56.

Horgan, Terrence. 1985. Newcomb's Problem: A Stalemate. In *Paradoxes of Rationality and Cooperation: Prisoner's Dilemma and Newcomb's Problem*, edited by R. Campbell and L. Snowden, pp. 223–34. Vancouver: University of British Columbia Press.

Jackson, Frank. 1979. On Assertion and Indicative Conditionals. *Philosophical Review* 88: 565–89.

Jaynes, E.T. 1978. Where Do We Stand on Maximum Entropy. In *The Maximum Entropy Formalism*, edited by R. Levine and M. Tribus, pp. 15–118. Cambridge, MA: MIT Press.

Jeffrey, Richard. 1968. Probable Knowledge. In *The Problem of Inductive Logic*, edited by Imre Lakatos, pp. 166–80. Amsterdam: North Holland.

Jeffrey, Richard. [1964]/1983. *The Logic of Decision*, 2nd edition. Chicago: University of Chicago Press.

Jeffrey, Richard. 1974. Frameworks for Preference. In *Essays on Economic Behavior under Uncertainty*, edited by M. Balch, D. McFadden, and S. Wu, pp. 74–9. Amsterdam: North Holland.

Jeffrey, Richard. 1977. A Note on the Kinematics of Preference. *Erkenntnis* 11: 135–41.

Jeffrey, Richard. 1978. Axiomatizing the Logic of Decision. In *Foundations and Applications of Decision Theory*, vol. 1, edited by C. Hooker, J. Leach and E. McClennen, pp. 227–31. Dordrecht: Reidel.

Jeffrey, Richard. 1983. Bayesianism with a Human Face. In *Testing Scientific Theories*, edited by J. Earman, *Midwest Studies in the Philosophy of Science*, vol. X, pp. 133–56. Minneapolis: University of Minnesota Press.

Jeffrey, Richard. 1987. Indefinite Probability Judgment: A Reply to Levi. *Philosophy of Science* 54: 586–91.

Jeffrey, Richard. 1991. Matter of Fact Conditionals. *Aristotelian Society* (suppl. vol.) 65: 161–83.

Joyce, James M. 1998. A Nonpragmatic Vindication of Probabilism. *Philosophy of Science* 65: 575–603.

Kaplan, Mark. 1983. Decision Theory as Philosophy. *Philosophy of Science* 50: 549–77.

Kaplan, Mark. 1989. Bayesianism without the Black Box. *Philosophy of Science* 56: 48–69.

Kaplan, Mark. 1996. *Decision Theory as Philosophy*. Cambridge: Cambridge University Press.

Kraft, C., J. Pratt, and A. Seidenberg. 1959. Intuitive Probability on Finite Sets. *Annals of Mathematical Statistics* 30: 408–19.

Kreps, David. 1988. Notes on the Theory of Choice. Boulder: Westview Press.

Levi, Isaac. 1980. *The Enterprise of Knowledge*. Cambridge, MA: MIT Press.

Levi, Isaac. 1982. A Note on Newcombmania. *Journal of Philosophy* 79: 377–82.

Levi, Isaac. 1985. Imprecision and Indeterminacy in Probability Judgment. *Philosophy of Science* 52: 390–409.

Levi, Isaac. 1986. The Paradoxes of Allais and Ellsberg. *Economics and Philosophy* 2: 23–53.

Lewis, David. 1973. *Counterfactuals*. Oxford: Blackwell.

Lewis, David. 1976. Probabilities of Conditionals and Conditional Probabilities. *Philosophical Review* 85: 297–315.

Lewis, David. 1979. Counterfactual Dependence and Time's Arrow. *Nous* 13: 455–76.

Lewis, David. 1980. A Subjectivist's Guide to Objective Chance. In *Studies in Inductive Logic and Probability*, edited by R. Jeffrey, vol. 2, pp. 263–94. Berkeley: University of California Press.

Lewis, David. 1981. Causal Decision Theory. *Australasian Journal of Philosophy* 59: 5–30.

Lewis, David. 1986. Probabilities of Conditionals and Conditional Probabilities II. *Philosophical Review* 95: 581–9.

Lewis, David. 1994. Chance and Credence: Humean Supervenience Debugged. *Mind* 103: 473–90.

Luce, R. Duncan and David Krantz. 1971. Conditional Expected Utility. *Econometrica* 39: 253–71.

Luce, R. Duncan and Howard Raiffa. 1957. *Games and Decisions: Introduction and Critical Survey.* New York: Wiley.

Machina, Mark. 1991. Dynamic Utility and Non-expected Utility. In *Foundations of Decision Theory*, edited by M. Bacharach and S. Hurley, pp. 39–91. Oxford: Basil Blackwell.

Maher, Patrick. 1993. *Betting on Theories.* Cambridge: Cambridge University Press.

McClennen, Edward. 1990. *Rationality and Dynamic Choice: Foundational Explorations.* Cambridge: Cambridge University Press.

McGee, Vann. 1994. Learning the Impossible. In *Probability and Conditionals*, edited by E. Eells and B. Skyrms, pp. 179–200. New York: Cambridge University Press.

Moser, Paul. 1989. *Knowledge and Evidence.* Cambridge: Cambridge University Press.

Nozick, Robert. 1969. Newcomb's Problem and Two Principles of Choice. In *Essays in Honor of Carl G. Hempel*, edited by N. Rescher, pp. 107–33. Dordrecht: Reidel.

Nozick, Robert. 1993. *The Nature of Rationality.* Princeton: Princeton University Press.

Pettit, Philip. 1991. Decision Theory and Folk Psychology. In *Foundations of Decision Theory*, edited by M. Bacharach and S. Hurley, pp. 147–75. Oxford: Basil Blackwell.

Popper, Karl. 1959. *The Logic of Scientific Discovery.* London: Hutchinson.

Price, Huw. 1986. Against Causal Decision Theory. *Synthese* 67: 195–212.

Price, Huw. 1991. Agency and Probabilistic Causality. *British Journal for the Philosophy of Science* 42: 157–76.

Price, Huw. 1993. The Direction of Causation: Ramsey's Ultimate Contingency. In *PSA 1992*, edited by D. Hull, M. Forbes, and K. Okruhlik, vol. 2, pp. 253–67. East Lansing: Philosophy of Science Association.

Ramsey, Frank. 1931. Truth and Probability. In *The Foundations of Mathematics and Other Logical Essays*, edited by R. Braithwaite, pp. 156–98. London: Kegan Paul.

Raz, Joseph. 1986. *The Morality of Freedom.* Oxford: Oxford University Press.

Reichenbach, Hans. 1959. *Modern Philosophy of Science.* London: Routledge and Kegan Paul.

Resnick, Michael. 1987. *Choices: An Introduction to Decision Theory.* Minneapolis: University of Minnesota Press.

Réyni, A. 1955. On a New Axiomatic Theory of Probability. *Acta Mathematica Academiae Scientiarium Hungaricae* 6: 285–335.

Royden, H.L. 1968. *Real Analysis.* New York: Macmillan Publishing.

Savage, Leonard. [1954]/1972. *The Foundations of Statistics*, 2nd edition New York: Dover.

Schervish, M., T. Seidenfeld, and J. Kadane. 1984. The Extent of Non-conglomerability in Finitely Additive Probabilities. *Zeitschrift fur Wahrscheinlichkeitstheorie* 66: 205–26.

Schervish, M., T. Seidenfeld, and J. Kadane. 1990. State-Dependent Utilities. *Journal of the American Statistical Association* 85: 840–7.

Scott, Dana. 1964. Measurement Structures and Linear Inequalities. *Journal of Mathematical Psychology* 1: 233–47.

Seidenfeld, Teddy. 1986. Entropy and Uncertainty. *Philosophy of Science* 53: 467–91.

Seidenfeld, T. and M. Schervish. 1983. A Conflict between Finite Additivity and Avoiding Dutch Book. *Philosophy of Science* 50: 398–412.

Shafir, Eldar and Tversky, Amos. 1995. Decision Making. In *Thinking*, edited by E. Smith and D. Osherson, pp. 77–100. Cambridge: MIT Press.

Shimony, Abner. 1994. Empirical and Rational Components in Scientific Confirmation. In *PSA 1994*, edited by D. Hull, M. Forbes, and R. Burian, vol. 2, pp. 146–55. East Lansing: Philosophy of Science Association.

Skyrms, Brian. 1980. *Causal Necessity*. New Haven: Yale University Press.

Skyrms, Brian. 1983. Three Ways to Give a Probability Assignment a Memory. In *Testing Scientific Theories,* edited by J. Earman, *Midwest Studies in the Philosophy of Science*, vol. X, pp. 157–62. Minneapolis: University of Minnesota Press.

Skyrms, Brian. 1984. *Pragmatics and Empiricism*. New Haven: Yale University Press.

Skyrms, Brian. 1987. Updating, Supposing, and Maxent. *Theory and Decision* 22: 225–46.

Skyrms, Brian. 1988. Conditional Chance. In *Probability and Causality*, edited by J. Fetzer, pp. 161–78. Dordrecht: Kluwer.

Skyrms, Brian. 1994. Adams Conditionals. In *Probability and Conditionals*, edited by E. Eells and B. Skyrms, pp. 13–26. New York: Cambridge University Press.

Sobel, Jordan Howard. 1994. *Taking Chances: Essays on Rational Choice*. New York: Cambridge University Press.

Spohn, Wolfgang. 1977. Where Luce and Krantz Really do Generalize Savage's Decision Model. *Erkenntnis* 11: 113–34.

Spohn, Wolfgang. 1986. The Representation of Popper Measures. *Topoi* 5: 69–74.

Stalnaker, Robert. 1968. A Theory of Conditionals. In *Studies in Logical Theory*, edited by N. Rescher, *American Philosophical Quarterly Monograph Series*, no. 2.

Stalnaker, Robert. 1968. *Inquiry*. Cambridge: Bradford Books.

Suppes, Patrick. 1970. *A Probabilistic Theory of Causality*. Amsterdam: North Holland.

Suppes, Patrick. 1974. *Probabilistic Metaphysics*. Uppsala: University of Uppsala Press.

Suppes, Patrick and Mario Zanotti. 1976. Necessary and Sufficient Conditions for the Existence of a Unique Measure Strictly Agreeing with a Qualitative Probability Ordering. *Journal of Philosophical Logic* 5: 431–8.

Teller, Paul. 1973. Conditionalization and Observation. *Synthese* 26: 218–58.

Todhunter, I. 1949. *A History of the Mathematical Theory of Probability*. New York: Chelsea Press.

Tversky, Amos. 1975. A Critique of Expected Utility Theory: Descriptive and Normative Considerations. *Erkenntnis* 9: 163–73.

van Fraassen, Bas. 1981. A Problem for Relative Information Minimizers in Probability Kinematics. *British Journal for the Philosophy of Science* 32: 375–79.

van Fraassen, Bas. 1984. Belief and the Will. *Journal of Philosophy* 81: 235–56.

van Fraassen, Bas. 1995. Fine-Grained Opinion, Probability, and the Logic of Belief. *Journal of Philosophical Logic* 24: 349–77.

Velleman, J. David. 1989. *Practical Reflection*. Princeton: Princeton University Press.

Villegas, C. 1964. On Qualitative Probability σ-Algebras. *Annals of Mathematical Statistics* 35: 1787–96.

von Neumann, John and Otto Morgenstern. 1953. *Theory of Games and Economic Behavior*, 3rd ed. Princeton: Princeton University Press.

Walton, Kendall. 1990. *Mimesis as Make-Believe*. Cambridge: Harvard University Press.

Index

Eells, Ellery, 154, 156–7, 165, 177n, 203n, 204n
efficacy value, 4, 161, 164, 175, 179, 224, 226, 228, 232
Ellsberg, Daniel, 101
Ellsberg's Paradox, 101–2
essential range, 237
event, 15, 66, 68; null, 86, 132
evidential decision theory (V-maximization), 3, 114–22, 127, 160, 164, 177–80, 224, 230–2, 252–3; non-uniqueness in, 122–7, 134–7, 146; answer to Newcomb's Paradox, 147–9, 160
evidential relevance (confirming power), 119, 203, 205–12
expectation, 17
expected utility, 3, 67, 78, 98, 136, 224; causal 3; evidential 3, 119–20; unconditional, 4, 116; conditional, 4, 119; causal, 161, 178

Fine, Terrence, 236
Fishburn, Peter, 29n, 96–7, 110n, 225–8, 226–8
Friedman, K., 217, 219

Gaifman, Haim, 166n
Garber, Daniel, 203n, 204n
Gardenfors, Peter, 183, 190, 191n, 197–8, 202, 214n
Gibbard, Allan, 150n, 167, 170–1, 173, 191n, 192n, 224–6
Glymour, Clark, 203
Godel, Kurt, 124n

Hajek, Alan, 38, 191, 192n
Hall, Ned, 191, 192n
Hammond, Peter, 201n, 214n
Hargreaves Heap, Shaun, 152n
Harper, William, 150, 167, 170–1, 173, 201n, 214n
Hellinger distance, 215, 217
Hitchcock, Christopher, 160n, 165
Hobson, A. 215n
Hofstadter, Douglas, 148n
Horgan, Terrence, 148n, 154n
Hume, David, 160
Huygens, Christian, 14n

imaging, 172, 174–6, 183, 196–9, 257; general, 198
incommensurable goods, 99–101
independence, evidential versus causal, 116–17
Independence Axiom, 27, 42, 85, 101–2

indifference, 69; contrasted with a lack of preference, 99–100
instrumental rationality (practical reasoning), 9, 111, 152, 160, 252
integral, Jeffrey, 124, 137

Jackson, Frank, 192
Jaynes, E. T., 215n, 216
Jeffrey, Richard, 4, 44n, 48, 49n, 50–2, 69–70, 102n, 103, 113, 114–15, 119–22, 124n, 126n, 127, 137n, 149, 154, 156–8, 177, 183, 203, 204n, 213, 252, 253; equation, 4, 119–22, 124, 133, 231–2, 242, 247; shift, 207, 213–15, 216, 223, 226
Jeffrey/Bolker Axioms, 127–33, 135, 137, 139, 233–4, 242; Nontriviality, 127; Partial Ordering, 127–128; Completeness, 128; Averaging, 128–9; Impartiality, 128–30; Coherence, 130–1, 136–6, 139; Nullity, 132; Null Consistency, 132; Archimedean Axiom, 132; Continuity, 132; Richness, 133

K-partition, 163–4, 170, 227
Kaplan, Mark, 43, 44n, 69n, 103n
Kashima, Yoshihisa, 102n
Kraft, C., 92, 256n
Krantz, David, 70, 108–10
Kreps, David, 28n, 29n, 38n
Kullback–Leibler information, 215, 217–19

learning, 206–9, 213, 216–19; nonmonotonic, 214
Levi, Isaac, 44, 81n, 102n, 214n
Lewis, David, 64n, 65n, 148n, 161, 163, 166n, 169–78, 186, 191–2, 196n, 197–8, 216, 257
likelihood ranking, 89–93, 130–1, 136, 138–9, 233–4; atomless, 92–3, 138; conditional, 189, 196, 200, 202, 229
Lindenbaum's Lemma, 103, 105
Luce, R. Duncan, 70, 98, 108–10

Machina, Mark, 53–4
Maher, Patrick, 97n, 102n
McClennen, Edward, 60n
McGee, Van, 201n
measurable function, 16, 184; support of, 17
mitigators, 109–10, 114, 228, 253
mixture, 23, 66
mixture space, 24
Moore, G. E., 87

Morgenstern, Otto, 23, 42, 78
Moser, Paul, 75n

Newcomb, William, 146n
Newcomb problem, 3, 146–8, 151–4, 156, 158–9, 170, 226; why aren't you rich objection, 151–4 news value (auspiciousness), 3, 119–20, 146, 149, 154, 157, 178–9, 181, 226, 231–2; of acts, 120–1
Nicole, Pierre, 9n, 10n
Nozick, Robert, 55n, 146n

old evidence, problem of, 200–15
Ordinal Uniqueness Theorem (for comparative probability), 134–5, 196, 237, 243
outcomes, 48–9; independence of, 53; underspecification of, 52–7; Jeffrey, 56, 68; Savage, 56; as unalloyed goods, 57; coarse grained, 68

partition invariance, 121–2, 176–8, 228
Pascal, Blaise, 10–19, 21, 42, 46, 78n
Pascal's Thesis, 18–23
Pascal's Wager, 37–8, 46, 94, 137
Pettit, Philip, 54n
plans, 58–9
Popper, Karl, 201n
pragmatism, 90, 131, 136, 226
Pratt, J., 92, 256n
preferences 40; representation of, 40, 98, 104; transitivity of, 84, 110; unconditional, 88, 106, conditional, 88, 106–7, 109; incompleteness of, 98–102, 127, not revealed by choice, 100–2; for news, 120, 130; preference ranking, 41, 68–9, 130; incomplete, 44, 98–102; conditional, 88, 106–7, 227–9, 237; unbounded 123, 126, 137
Price, Huw, 154, 156, 159–61, 179
Principal Principle, 166–7, 173n
Prisoner's Dilemma (with a twin), 148
probability, 4, 11; conditional, 4, 107, 115, 119, 160–1, 185, 190, 195–6, 201–203, 204, 232, 236–7; comparative, 91, 130–1, 134–6, 183, 234; uniqueness of representations, 93, 133–9; atomless, 134; causal, 161–72
Problem of the Points, 10–14
proposition, 50, 78
pure rationality, axioms of, 82, 97, 103, 128

Raiffa, Howard, 98

Ramsey, Frank, 23, 42n, 78, 189
Ramsey test, 189–93; weak, 193–4, 199, 202
ratifiablility, 154–9, 226; evidential, 158
Raz, Joseph, 99n
regularity, 188, 198, 220
Reichenbach, Hans, 155–6
relevance, causal versus evidential, 119, 162–3
representation, of preferences, 41, 45, 98; of beliefs, 91–3, 106
representation theorems, 4, 78–82; Savage's, 96, 127, 103n, 112; Bolker's, 127, 133; Bolker's Theorem, generalized, 138–45
Réyni–Popper measures, 164n, 200–15, 222–3, 229, 231–2
risk, 16; versus uncertainty, 16; aversion to, 35; seeking, 35, neutrality, 31
Royden, H., 17n

St. Petersburg Paradox, 32, 37–8, 137
Salmon, Wesley, 155n
Savage, Leonard, 6, 15, 18n, 23, 28n, 29n, 48–52, 50n, 59–60, 69n, 69–70, 73, 77, 80n, 85, 96, 111n, 113n, 114, 131, 133, 134, 176–7, 234n; equation, 79, 111–12, 115, 118–19, 164, 226; representation theorem, 96, 127, 225–6; on independence of acts and states, 115–18;
Savage's Axioms, 83–95; Richness, 83, 106; Nontriviality, 83; Partial Ordering, 84; Completeness, 84, 98; Independence, 85, 115, 128, 130; Nullity, 87, 115, 128; Stochastic Dominance, 88; Coherence, 89; Event Richness, 93, 105–6; Averaging, 95; Dominance 97; status of, 97–110
scale, 29; absolute, 122; interval, 29, 122–3, 126
Schervish, M., 53n, 89, 90n, 96n
Scott, Dana, 256
screening off, 155–6, 158
sections, 238
sectional representation, 233–5, 237
Seidenfeld, Teddy, 89, 90n, 215n
Seidenberg, A., 92, 256n
selection rule, 184, 215, 216
Shafir, Eldar, 14n, 94
Shimony, Abner, 204, 217, 219
similarity gauge, 185, 215, 220
Skyrms, Brian, 102n, 163n, 164, 166n, 167n, 171, 184n, 186, 209n, 215, 216–18

267